BRANCHING OUT

L. E. MAKAROFF

First Edition

Copyright © 2024 by LE Makaroff

Cover Design: T Staskevich

❀ Created with Vellum

For my grandmother, Gwen

CONTENTS

PART THREE

PREFACE

Do not be afraid to take chances or make mistakes. That is how you grow. Pain nourishes your courage. You have to fail in order to practise being brave.

MARY TYLER MOORE

So, you've finally accepted that you will finish that Master's or PhD. Your whole life, you have expected to continue ticking off the boxes in academia:

- Bachelor degree
- Master's degree
- PhD
- Post-doc
- Tenure-track professor
- Tenured professor

But now you're not so sure. If you've realised maybe academia isn't for you, this book will help you find your way out.

You have not failed by leaving academia.

The great news is that a master's or PhD opens so many doors to you – you just need to know where to look. After going through this process myself, I will reveal how to find and open these doors and explore some of the many paths away from academia.

I'm excited to share a bit of my story with you and, hopefully, give you some handy tips along the way.

My adventures in life sciences started at the Australian National University, studying for both a Bachelor of Science and a Bachelor of Psychology. These degrees opened my eyes to the amazing world of medical research and the human side of science. I went on to do a PhD in Medical Research and a Master of Public Health. Those years were about understanding the many facets of life sciences.

My career has seen me take on a variety of roles across different sectors.

I worked as a Senior Post-Doctoral Research Fellow at an American university where breakthrough discoveries are made. I was right in the thick of academic research, getting my hands dirty with real-world science problems.

I have worked at an international consultancy that supports the life sciences industry. As a Research Consultant, I got to dip my toes in different projects and learned a lot about the business side of science.

I had a stint at a big global pharmaceutical company. My job there as a Health Outcomes Manager was a real eye-opener. I learnt about the pharmaceutical industry and how our work can directly help people with severe health conditions.

Working with umbrella organisations has taught me about bringing different voices together – from doctors and nurses to patients – to make healthcare better for everyone.

As Chief Executive and President of cancer organisations, I've had the opportunity to team up with strong personalities, learning the value of cooperation and prioritising the needs of those we serve. These jobs have shown me how science can have a real impact on people's lives.

All these experiences have been more than just jobs to me. They've been adventures, learning curves, and a chance to make a bit of a difference in the world. I hope by sharing my stories and insights, I can help you find your own path in the crazy, wonderful world of life sciences.

In this book, I'll share the good, the bad, and the lessons learned from both. My journey's had its share of bumps and wins, and I hope my experiences can help guide and inspire you as you start your own path in life sciences.

I chose to write this book because I receive so many messages like this, and I wanted to help:

> After ten years in a lab, I am looking to change careers. I don't really have a clue of what to pursue next. I feel like working as a research assistant hasn't given me many transferable skills. I think I'd enjoy something to do with events around life science, and I'd love a job where I could travel. I wanted to ask if you could think of any tips, or had any recommendations or contacts, but mainly I would be so grateful if you could have a look at my CV/cover letter as I feel like I don't have a clue of how to make it stand out."

> I am looking to move away from academic research. I think it would still be beneficial to discuss the transition process as it does feel relatively more daunting looking for employment

outside of the academic world. I am still trying to work out how to best approach making applications to charitable organisations."

What did I love about my PhD? Well, the lab was my second home. Or should I say first home, as I indeed spent more time in the lab than anywhere else, including my residential address.

The other PhD students I worked with were incredible, and we developed a bond that can only be created when you spend 24 hours together, seven days a week. We would finish up each day of lab work with a microwave dinner in the kitchenette, and then we would review each other's presentations and documents. We dried each other's tears after yet another failed experiment, yet another unreasonable supervisor or yet another general rejection, and then we would either (depending on our energy levels) go out drinking and dancing or get into our pyjamas and have a movie night.

There is nothing like shared adversity to strengthen bonds; luckily, the mind has a pleasant way of smoothing over the rough edges of rejection and despair and retaining the warmth and joy of fellowship.

Interestingly, of all the lab mates that I went through my PhD with, none of them became professors. But they all achieved professional success and personal happiness.

One of them went on to do a medical degree after her PhD; now she is a very successful anaesthesiologist.

One of them went on to work in a government laboratory to ensure public health regarding contagious diseases and is now responsible for the public health of millions of citizens.

One decided that her passion was linguistics and took up a part-time job as a technician while learning linguistics, raising children and dogs, and investing in real estate.

There are many paths to success. True success is happiness. It is lying on your deathbed, looking back at your life, and feeling a sense of satisfaction.

While I'm glad I am no longer in academia, I am happy for everything it has taught me.

I am not afraid of scientific papers, except those regarding the hadron particle. But anything else, I know I can decipher. Scientific words are a language that I have mastered.

I know how to analyse evidence. I know the difference between an anecdote and a randomised clinical trial. I can understand new medical treatments.

Why did I decide to leave academia? Well, that is a good question. Which incidentally is my go-to thinking phrase whenever I am asked a tricky question during an interview or an oral presentation. One reason is that I struggled to master lab skills. I cannot bake, I cannot draw, I cannot garden, and therefore I should not have been surprised that I also derive very little joy from transferring one small piece of liquid from one container into another container. Having to pour gels, run PCRs, wake up at 4 am to run cells through a flow cytometry machine, and ensure that everything was constantly sterile... none of these tasks were for me. I discovered I could be meticulous when it comes to a spreadsheet, but not when it comes to my lab bench.

Sometimes I wonder if I would've stayed in academia if my work had been purely theoretical. If I had spent all my days in terms of theoretical epidemiology or quantum physics or statistical analysis of wave fluctuations, maybe I would've stuck around. Or maybe not.

In academia, there is constant pressure to publish or perish. Every day not only do you need to have a complete understanding of the field of research in which you are working, but you must be able to take a bird's eye view to see the gaps, scan the horizon, and know the direction you want to walk towards. Then you also need to have scientific communication skills to write the papers, give international keynote presentations, and get those grants. You also need to be able to handle interview panels, not only for your grants but also for those faculty positions.

The most straightforward answer for why I left academia is that it just wasn't bringing me joy. I just wasn't a good match. Eventually, I learned that is enough of an answer. You don't need to justify leaving academia. If you wake up in the morning and are not excited to go to work, and this lack of excitement has been going on for more than six months, then maybe this isn't the right career path for you.

One of the most challenging aspects of leaving academia is this feeling of failure. There is this unspoken idea that to be a success, one must continue to the pinnacle of receiving the title of "Professor". To deviate from this path is to fail. It is to give up. It is to quit.

It can be difficult to differentiate between stopping during a PhD and leaving after a PhD. Once you have that "Dr" in front of your name and that beautiful "PhD" after your name, there is no harm in leaving. And there might be harm in staying.

This book is going to help you find your path. Not only will I help you identify your strengths and limitations and find the ideal career path for you, but I will also help you move into that career path with templates for cover letters, CVs, interview answers, LinkedIn, X (formerly Twitter), Mastodon, and follow-up emails.

You've got this. Whether you want to escape from academia or thrive within academia, this is your tool kit to do just that.

SUMMARY

- Leaving academia is not a failure; it's a valid decision that opens up new opportunities.
- A master's or PhD degree offers numerous career paths outside of academia that should be explored.
- Success is defined by personal happiness and satisfaction, and there are various paths to achieve it outside of traditional academic roles.

PART ONE

PART ONE

NAVIGATING THE PATH LESS TRAVELLED

Life presents us with countless paths. Some are well-trodden and clearly marked, offering a sense of security and predictability. Others, less defined, weave through uncharted territories, brimming with mystery and potential. How does one choose a path when the destinations are manifold and the signposts, at times, enigmatic?

In this section, I introduce a quiz designed to serve as a compass, guiding you through the maze of career choices. You will explore what drives you, what work environment resonates with your nature, the tasks that bring you joy, and the memories that shape your academic journey.

As you navigate through the questions, you will be encouraged to think about your ideal work week, your approach to problem-solving, your preferred modes of communication, and what motivates you in a job.

Upon completing the quiz, you will find a section aligning your responses with potential career paths. This is not a definitive answer but, an invitation to consider possibilities that resonate with your personality and aspirations.

CHAPTER 1
QUIZ

*Alice: Would you tell me, please, which way I ought to
    go from here?*
*The Cheshire Cat: That depends a good deal on where
    you want to get to.*
Alice: I don't much care where.
*The Cheshire Cat: Then it doesn't much matter which
    way you go.*
Alice: ...So long as I get somewhere.
*The Cheshire Cat: Oh, you're sure to do that, if only you
    walk long enough.*

All your life you've had a clear roadmap:

High School -> Undergraduate -> Graduate -> Post-doc

It's time for you to use this quiz to stop and reflect. You can also take this quiz at

www.makaroff.ink

What do you want from your life? Tick all that apply:

A. A job I love

B. Making a significant societal impact

C. Having fun

D. Never any overtime

E. Working with cutting-edge technology

F. A clear structure

G. Freedom and autonomy

H. An impressive job title

I. A high salary

J. Travelling the world

K. Providing care and support

L. Minimal stress

M. Influencing future generations

N. Knowledge

What type of work environment do you prefer? Tick all that apply:

A. Collaborative and supportive

B. Innovative and pioneering

C. Historic and inspiring

D. Structured and stable

E. Safety-conscious and organised

F. Detail-oriented and technical

G. Creative and expressive

H. Research-focused and analytical

I. Independent and self-guided

J. Fast-paced and dynamic

K. Empathetic and caring

L. Logical and methodical

M. Educational and nurturing

N. Intellectual and academic

What type of work tasks do you enjoy most? Tick all that apply:

A. Helping others and making a difference

B. Innovating and implementing new ideas

C. Managing projects and people

D. Organizing and streamlining processes

E. Ensuring safety and compliance

F. Performing technical tasks with precision

G. Creatively expressing ideas and concepts

H. Conducting research and analysis

I. Strategic planning and decision-making

J. Analysing data and problem-solving

K. Providing care and support

L. Working with numbers and logic

M. Educating and inspiring students

N. Pursuing and publishing research

What was your favourite part of undergraduate studies? Tick all that apply:

A. Participating in community service

B. Managing a club or society

C. Studying history

D. Getting involved in clubs focused on policy

E. Developing practical skills

F. Gaining hands-on equipment experience

G. The focus on writing

H. The focus on business

I. Sitting exams

J. Networking with faculty

K. The focus on the human body

L. The focus on statistics

M. Teaching assistantships

N. Academic research opportunities

What are your strengths? Tick all that apply:

A. Ability to write over 20 emails a day

B. Creative problem-solving skills

C. Focus

D. Highly organised and detail-oriented

E. Methodical and precise

F. Excellent hand-eye co-ordination

G. Passion for writing

H. Enjoy giving presentations

I. Skilled in brainstorming

J. Love for face-to-face meetings

K. Can make decisions quickly

L. Enjoy working with data

M. Teaching and mentoring

N. Ability to withstand critiques from peers

What do you remember most easily? Tick all that apply

A. Personal stories

B. Procedures

C. Artistic concepts

D. Historical events

E. Case studies

F. How things work

G. Vivid descriptions

G. Presentations

H. Patterns

I. Ted talks

J. Strategy details

K. Anatomy and physiology

L. Trends

M. Faces

N. Concepts

What has been the best part of your graduate experience?: Tick all that apply:

A. Working with my colleagues

B. Mentoring undergraduates

C. Engaging in interdisciplinary projects

D. Having a structure to my days

E. Organising my bench and the lab

F. Discovering something new

G. Writing up my thesis

H. Giving presentations

I. Implementing new methods

J. Travelling to conferences

K. Learning more about the human body

L. Crunching the numbers

M. Curriculum development

N. Academic research and publishing

What do you want your work week to look like? Tick all that apply:

A. Every day is different

B. Regularly collaborating with a team

C. Focused on creative projects

D. Working from the office

E. Start at 9, leave at 5, minimal travel

F. Days full of experiments

G. Working from home with few meetings

H. Working from the office with meetings

I. Constant communication with colleagues

J. Every week I'm in a different city

K. Days full of face-to-face interactions

L. Days full of analysis

M. Days full of face-to-face interactions

N. Research and lectures

How do you approach problem-solving? Tick all that apply:

A. I prefer a people-oriented approach

B. I formulate strategic plans

C. I delve into cultural knowledge for insights

D. I like to follow established procedures

E. I focus on risk assessment

F. I enjoy hands-on and practical methods

G. I use writing to explore solutions

H. I think about scientific approaches

I. I think about innovative approaches

J. I enjoy brainstorming

K. I consider the ethical implications

L. I rely on statistical analysis

M. I seek pedagogical insights

N. I approach it theoretically

How do you prefer to communicate in a professional setting? Tick all that apply:

A. Personal conversations

B. Email correspondence

C. Collaborative team meetings

D. Formal reports

E. Briefings

F. Hands-on demonstrations

G. Writing reports

H. Scientific presentations

I. Business meetings

J. Presentations and pitches

K. Patient consultations and medical histories

L. Data visualisations and analytical reports

M. Classroom teaching

N. Lectures and academic papers

What motivates you most in a job? Tick all that apply:

A. Making a positive impact in the community

B. Creating innovative solutions

C. Creating a legacy

D. Stability and structure

E. Ensuring safety and compliance

F. Solving technical challenges

G. Communicating complex information

H. Mentoring and managing teams

I. Being at the forefront of scientific innovation

J. Opportunities to network

K. Improving others' health and well-being

L. Uncovering insights through data analysis

M. Teaching others

N. Discovering new knowledge

What is your preferred method of learning? Tick all that apply:

A. By engaging with people

B. Through creative exploration

C. By exploring historical contexts

D. Through structured courses

E. By learning about regulations

F. Through hands-on practice

G. By reading and writing

H. Through research

I. By testing new technologies

J. Through interactive workshops

K. Through seminars

L. Through data analysis

M. Through educational practice

N. Through academic study

Imagine you're on your deathbed at 96, looking back at your life. What are you most proud of? Tick all that apply:

A. I left a legacy that will help people

B. I nurtured successful teams

C. I shared cultural heritage

D. I made a contribution to society

E. I protected my colleagues

F. The technical challenges I overcame

G. I helped people learn about the world

H. I helped develop life-saving treatments

I. I pushed the boundaries of technology

J. I travelled the world

K. The comfort I provided

L. The insights I uncovered

M. I inspired imagination

N. The research I published

What is your long-term career goal? Tick all that apply:

A. Making a positive impact on society

B. Championing diversity and inclusion

C. Preserving and sharing cultural heritage

D. Contributing to public service

E. Ensuring safety in the workplace

F. Excelling in a technical or specialised field

G. Expressing creativity through writing

H. Being part of groundbreaking research

I. Leading in business development

J. Achieving leadership in a corporate setting

K. Providing care and support to others

L. Utilising data analysis to drive decisions

M. Inspiring others through education

N. Advancing knowledge

How would your friends describe you? Tick all that apply:

A. Open-minded

B. Compassionate

C. Talented

D. Reliable

E. Detail-oriented

F. Creative

G. Inquisitive

H. Friendly

I. Adaptable

J. Ambitious

K. Patient

L. Analytical

M. Inspirational

N. Intellectual

What do you hate most? Tick all that apply:

A. Apathy

B. Inefficiency

C. Lack of creativity

D. Corruption

E. Ignorance

F. Poorly designed systems

G. Misinformation

H. Isolation

I. Stagnation

J. Missing out

K. Lack of empathy

L. Interruptions

M. Injustice

N. Dishonesty

Now, let's see where your answers might lead you in terms of career paths. This part of the quiz aligns your most frequent choices with potential fields of work, offering a glimpse into where your natural inclinations could thrive.

If you got:

Mostly "As" – you could consider working in a small charity

Mostly "B"s – you could consider working in a large charity

Mostly "C"s – you could consider working in a museum

Mostly "D"s – you could consider working in the civil service

Mostly "E"s – you could consider working as a health and safety officer

Mostly "F"s – you could consider working as a technician

Mostly "G"s – you could consider working as a medical writer or journalist

Mostly "H"s – you could consider working in a pharmaceutical company

Mostly "I"s – you could consider working in a biotech start-up

Mostly "J"s – you could consider working in a consultancy

Mostly "K"s – you could consider working in medicine

Mostly "L"s – you could consider working as a statistician or clinical data manager

Mostly "M"s – you could consider working as a teacher

Mostly "N"s – you could consider working as a tenure-track professor

Now, let's pause.

How do these results sit with you? Are you surprised by the career path suggested? Perhaps it aligns perfectly with your interests and strengths, or maybe it's opened up a new perspective on a field you hadn't considered. Remember, these suggestions are based on your quiz responses, reflecting aspects of your personality and preferences. Whether you find them spot-on or unexpected, use this insight as a starting point for further exploration. Your ideal career path is one that not only suits your skills but also resonates with your passions and values

CHAPTER 2
ADVICE FOR ALL

Done is better than perfect.

SHERYL
SANDBERG

PERFECT IS THE ENEMY OF GOOD

Making a career change can be daunting, especially if you feel like you are stuck in a job that doesn't fulfil you. It's easy to get trapped in a cycle of waiting for the perfect opportunity to come along, but the truth is that there is no such thing as the perfect opportunity. Waiting for everything to align before making a change can cause you to miss out on opportunities that could have helped you reach your goals.

If you're considering a career change, it's important to start taking action as soon as possible. Begin by researching the industries and companies that interest you. Reach out to people in those fields for informational conversations to learn more about the day-to-day realities of their jobs. You might even consider taking a class or workshop to gain new skills or knowledge that can help you transition into your desired field.

While it's important to act, it's also important to be realistic about your goals and expectations. Don't put too much pressure on yourself to have everything figured out right away. Instead, focus on taking small steps towards your goal and adjust your path as needed. Remember, it's better to start somewhere than to never start at all.

When considering a career change, it's important to keep in mind that it's best not to quit your job until you have a new one lined up. Being unemployed can often be seen as a red flag by potential employers. So, while it's good to take action, be sure to do so in a calculated and strategic way. Take the time to research your options and build your network before making any big moves.

IT'S NEVER TOO LATE

It's never too late to pursue your dreams, even if you feel like you may have missed the boat. Don't compare yourself to others and their timelines of success, as everyone's journey is unique. Remember that you have the power to change your career path at any age. While it may seem daunting to start over or transition to a new field, it's never too late to chase after what you truly want in life.

You may feel like you're behind or that it will take a long time to reach your goals, but it's important to remember that time will pass regardless. So, why not use that time to work towards your dream job or career? Whether it takes a year, five years, or even ten years, the time will keep moving. Embrace the journey and trust in your ability to succeed, no matter your age or previous experience.

BE FLEXIBLE

As you progress in your career, it's important to remain open to change and adapt to new circumstances. It's natural to experience shifts in your interests, priorities, and values over time. What you thought you wanted when you first entered the workforce may no longer align with your current goals and aspirations.

Changing your career focus does not necessarily mean starting over from scratch. You may be able to transfer some of the skills and experiences you've gained in your current career to a new one. Take some time to assess your strengths, interests, and values, and look for areas where they overlap with potential new career paths. By staying flexible and open-minded, you can find a new career path that better aligns with your goals and aspirations, and ultimately lead a more fulfilling professional life.

MOVE OVERSEAS

Moving overseas is a unique opportunity that not many people have the privilege to experience. It's a chance to challenge yourself, explore new places, and gain invaluable life experiences. Whether it's for work or personal reasons, moving overseas is a decision that requires careful consideration and planning. However, if you have the means and opportunity to do so, it may be worth seizing the moment. You may find that your dream job or career path is waiting for you in a new country. By expanding your geographical scope, you open yourself up to endless possibilities and opportunities.

As someone who has lived in four countries, I can tell you that one of the most exciting aspects of moving overseas is the opportunity to discover new cultures and lifestyles. Every weekend can feel like a holiday as you explore your new city, try interesting foods, and learn about different customs and traditions. It can be an eye-opening experience that helps you grow as a person and gain a new perspective on life.

It's natural to feel apprehensive about taking risks and stepping out of your comfort zone. However, moving overseas can be a powerful catalyst for personal growth and development. You may discover hidden strengths and abilities that you never knew you had, and you'll gain the confidence to take on even more challenges. Additionally, you'll learn to navigate new environments and situations and develop a greater appreciation for diversity and multiculturalism.

Of course, moving overseas can also be challenging, and it's important to prepare yourself for potential difficulties such as language barriers, homesickness, and culture shock. However, with the right attitude and support network, you can overcome these obstacles and thrive in your new environment. Ultimately, the benefits of moving overseas are many and varied, and it's an experience that can positively impact your life in countless ways. So if you have the opportunity to move abroad, go for it and embrace the adventure!

~

SUMMARY

- Start taking action in your career change journey, researching industries and companies, connecting with professionals, and gaining new skills.
- Remember that it's never too late to pursue your dreams and change your career path. Embrace the journey and trust in your ability to succeed.
- Stay flexible and open to change, assessing your strengths, interests, and values to find a new career path that aligns with your goals.

PART TWO

PART TWO

YOUR POTENTIAL FUTURES

The next section is designed to provide you with a more comprehensive understanding of each career option.

In the following pages, we'll embark on a journey through each potential career path. For each profession, we'll uncover what the day-to-day looks like, the skills and qualities that are essential for success, and the potential challenges and rewards you might encounter.

As you read through each section, I encourage you to reflect on your abilities and qualifications, and also on what ignites your passion and drives your curiosity. Keep an open mind; sometimes the best career path is the one you least expect.

The journey to finding the right career is unique for everyone. It's perfectly okay if you resonate with more than one option or if none of the suggested paths feel quite right. The final decision is yours, based on your own experiences and aspirations.

So, let's turn the page and begin our in-depth exploration of each potential career option. Your future awaits, rich with possibilities and opportunities for growth and fulfilment. Here's to finding the path that's just right for you.

CHAPTER 3
SMALL CHARITY

To laugh often and much; to win the respect of the intelligent people and the affection of children; to earn the appreciation of honest critics and endure the betrayal of false friends; to appreciate beauty; to find the best in others; to leave the world a bit better whether by a healthy child, a garden patch, or a redeemed social condition; to know that one life has breathed easier because you lived here. This is to have succeeded.

RALPH WALDO EMERSON

ended up in the charity sector, so obviously, I love it. For me, it's the best of all worlds. We are a small team of seven staff, a dozen trustees, and a handful of volunteers. I know everyone personally.

I have freedom, autonomy, mastery, and purpose.

Why do I love working at a charity? Well, besides those benefits I just mentioned, there are so many reasons. Firstly, some days it feels more like a hobby than a job. I love it so much. Every day, I get to do something different, and I get to do something meaningful. It makes me feel good when I tell someone I work for a charity. I can look in the mirror and know I am contributing to making the world a better place.

Jess, a colleague working in a different charity, says this about her experience:

> "I love being able to speak with scientists, communities, and policymakers all in the same week, and to join some of those dots to bring science into the real world, and in turn to feedback which science is most needed to support calls for policies to improve lives. I also love working with others who are trying to make the world, or part of it, a better place. They tend to be outward looking, selfless, and inspiring."

Aspects of a job like this make us feel more like a human and less like a cog in a big machine.

I work with a small team, but they are all amazing. There are minimal layers of bureaucracy to getting something done. The board sets the vision, and then we work towards it. We meet people every day whose lives have been made better by our work. My work touches many different areas. I can work with policymakers, I can work with doctors, I can work with politicians.

My job allows me to travel to new places to conferences to listen to what's happening around the world and learn from that.

And I can make decisions quickly.

How do you know if working for a charity is for you? Well, you need to be willing to step into the unknown. You had to do this during your PhD or post-doc when your supervisor asked you to figure out a new protocol, or you had to write your first scientific paper. But there is not much of a safety net in a charity. There's no one else to help you a lot of the time.

Even as a Chief Executive, I'm still cleaning the cups in the kitchen, and I'm still diagnosing video conference problems, but I love the fact of a new challenge every day. You need technical competence to be in a charity in the 21st century. So much of what charities do is reliant on technology, and you're not going to have anyone else to set it up for you. So you do need to have quite a high level of technical literacy.

ADVANTAGES OF WORKING IN A SMALL CHARITY

- Driven by altruism rather than by profit margins
- Agile due to small team and responsive leadership
- You know everyone in your organisation
- Usually, a clear mission and purpose
- Engagement with a diversity of stakeholders

DISADVANTAGES OF WORKING IN A SMALL CHARITY

- A dysfunctional board can encourage a toxic work environment
- The job is never done; the problem you're addressing is never completely "fixed"
- Often, you must be a jack of all trades
- You might have to do a lot of grunt work
- It can be difficult to be accountable to a board who doesn't work with you daily
- The pay is typically lower than the private sector

FILL THE GAPS IN YOUR CV ON YOUR OWN

- Follow the news related to issues you're interested in
- Follow updates from charities that inspire you
- Volunteer for a charity

FILL THE GAPS IN YOUR CV IN YOUR WORK

- Volunteer to work on a strategy committee
- Get involved with a charity that funds your research
- Enter competitions for posters and presentations
- Write a short monthly report to your supervisor
- Take a course in presentation skills
- Take a class in Excel skills
- Start a X (formerly Twitter) or Mastodon account
- Ask to draft a press release
- Contribute to the newsletters and social media accounts
- Ask to organise a meeting with non-academic partners
- Travel to congresses and give oral presentations
- Ask to manage volunteers and students
- Apply for grants

GREEN FLAGS IN JOB DESCRIPTIONS

- Staff members are visible on the website
- Staff members are celebrated on social media
- Salary is mentioned
- Generous holiday allowance
- Additional pension contributions are stated
- Remote working or hybrid working is encouraged
- Flexible working hours

RED FLAGS IN JOB DESCRIPTIONS

- Multiple vacancies are advertised for a small organisation
- Only board members on the website, not staff members
- Salary is not mentioned
- "ability to work under pressure"
- "young team"
- "vacation to be taken according to the needs of the Board"
- "ensure all projects operate flawlessly"
- "availability to undertake business travel at short notice"
- "willingness to work longer hours when needed"
- "very involved and expert Board"
- "character traits of being flexible and stress-resistant"

THE BEST COURSES TO TAKE

Non-profit Organisations

`coursera.org/learn/nonprofit-organizations`

This course, delivered remotely by the University at Buffalo, provides learners with a foundational understanding of the non-profit sector, offering insights and case-studies in leadership, governance, and what it means to be a board member. This course is self-paced, however, it usually takes five weeks to complete. Participants will explore the unique aspects of non-profits, as well as the importance of good governance, the responsibilities of board members, and the responsibilities of staff members. They will also engage in assessments to apply their knowledge. A certificate can be earned and shared on your LinkedIn profile.

Program Evaluation for Your Non-Profit

`coursera.org/projects/program-evaluation-non-profit`

This 2-hour guided project is designed to teach participants how to create a program evaluation plan for non-profits. The course covers the importance of program evaluation, the use of Logic Models, writing SMART goals, and formulating effective evaluation questions. It provides skill-based, hands-on learning with expert guidance and a shareable certificate upon completion.

~

SAMPLE COVER LETTER

Dear Mrs Tova,

I am excited about the opportunity for the Health Policy and Strategy Manager position at your umbrella charity. As a PhD graduate student with a focus on health and strategy, I am eager to apply my academic knowledge and research skills to this role.

I have engaged in various group projects, which have honed my ability to collaborate effectively with peers. I led a university team on a research project, which received a commendation for its innovative approach. My academic journey has also involved presenting research findings to diverse audiences, strengthening my public communication skills.

My academic background provides a solid foundation in analysis and strategy development. I have a strong understanding of systems and policy frameworks, which would be valuable in your organisation.

I have also had the opportunity to interact with healthcare professionals and policymakers, helping me to build a network in the health policy field. My experience in these settings has given me insight into cross-cultural collaboration and the nuances of policy-making processes.

I am excited about the possibility of bringing my academic insights and fresh perspective to your team. I am available for an interview throughout August and am keen to discuss how I can contribute to the success of your charity.

Thank you for considering my application.

Sincerely,

Dr Sam Tang

~

SAMPLE CV

Dr Sam TANG, MPH PhD

Experienced English-speaking PhD graduate, skilled in team motivation, relationship-building, and quality delivery.

EXPERIENCE & EDUCATION

Graduate Research Assistant

Banting University, Osgiliath, Gondor

- Implemented project management techniques that reduced project completion time by 15%.
- Authored 3 significant research papers published in top-tier journals, cited over 50 times collectively.
- Contributed to 4 innovative grant applications, which resulted in a 20% increase in research funding for the lab.

PhD in Immunology

Banting University, Osgiliath, Gondor

- Successfully led a research team of 8 members in a groundbreaking immunology project.
- Demonstrated project management by coordinating laboratory resources, resulting in the project's completion 2 months ahead of schedule.
- Developed and implemented a strategic project plan that increased lab efficiency by 35%.

Public Policy Challenges of the 21st Century

University of Virginia, Eddis

- Conducted a policy analysis project that was selected for presentation.
- Collaborated on a community engagement initiative.
- Authored a white paper on sustainable development policies

Masters of Public Health

Worthington University, Gondor

- Specialised in Epidemiology and Health Policy, ranked in the top 10% of the class.
- Led a public health campaign, reaching 300 local residents with preventative health information.
- Organised a 4-part seminar series on public health challenges and solutions, attracting over 200 attendees.

Statistical Methods for Research in the Life Sciences

Banting University, Gondor

- Developed a statistical model that improved data analysis efficiency by 40%.
- Designed a survey for a large-scale epidemiological study involving 1,200 participants, ensuring data quality and reliability.
- Contributed to a research team studying the efficacy of 5 different public health interventions.

SELECTED PUBLICATIONS

Tesfaye K, Nguyen L, Ilunga U, Deng J, Cho NH, Tang LE. International Health. 2115. Pyrus: International Health Coalition

Tesfaye S, Jalang'o A, Gandhi N, Hill L, Tang LE et al. Immunology: A review. 2115. Pyrus: International Health Coalition

Tesfaye BY, Dodd S, Scott C, Jacqmain O, Wilson P, Tang LE, et al. Global Health Scorecard: tracking progress for action. 2114. Pyrus: International Health Coalition

Tang LE and Ilunga U. The burden of disease. Health Voice. 1014. 59(4):35-39

HOBBIES

Metal detecting, fossil hunting, gem cutting

∾

L. E. MAKAROFF

SUMMARY

- **Working in a small charity offers freedom, autonomy, and a sense of purpose, with opportunities to make a meaningful impact and engage with diverse stakeholders.**
- **Advantages of small charities include a driven altruistic focus, agility due to a small team and responsive leadership, and a close-knit working environment where everyone is known personally.**
- **Disadvantages may include a potential for toxic work environments, a never-ending workload, the need to be versatile, and relatively lower pay.**

CHAPTER 4
LARGE CHARITY

You make a living by what you get. You make a life by what you give.

WINSTON CHURCHILL

Working in a large charity is a unique and fulfilling experience. It's a blend of passion and professionalism, where you contribute to societal improvement while engaging with a diverse array of stakeholders. In a large charity, the scope of influence is broad, and the impact is significant.

In a large charity, you're part of a bigger team compared to smaller organisations. While this might introduce more layers of bureaucracy, it also means more specialised roles and resources. The vision set by the charity guides your efforts, leading to tangible improvements in people's lives. Your role could span various domains – from policy advocacy to collaborating with medical professionals and political figures.

Travel is often part of the job, attending conferences and learning about global developments. Decision-making can be swift, though sometimes it requires navigating through more complex organisational structures.

Technical competence is crucial in today's charity sector. This includes expertise in Customer Relationship Management systems like Salesforce for donor and volunteer management, data analysis tools such as Excel and Tableau for evaluating fundraising efforts, digital marketing platforms like Mailchimp and Google Analytics for online engagement, Content Management Systems like WordPress for website upkeep, and project management tools like Trello for organising initiatives. Financial management software, online fundraising platforms, cybersecurity awareness, cloud computing services, and social media proficiency are also vital, supporting charities in financial handling, fundraising, data protection, resource management, and stakeholder engagement. Most of these skills can be learned through free online classes, either immediately before your job begins, or as on-the-job training.

One of the primary issues faced in this sector is the potential for bureaucratic complexities, often due to the large scale and scope of operations, which can lead to slower decision-making processes and administrative hurdles. Additionally, the nature of work in charities tends to be relentless, driven by the ongoing need to address social, environmental, or humanitarian issues that rarely have straightforward or quick solutions. This demands a high level of commitment and resilience from those working in the sector. Moreover, the job often requires a broad skill set to effectively meet the diverse needs of the charity's objectives. Typically, salaries and financial rewards in large charities are lower than those in the private sector, reflecting the altruistic nature of the work rather than a focus on profit. This aspect can be particularly challenging for people who are balancing a passion for their cause with practical financial needs.

The suitability for charity work mirrors the explorative nature of academic research, like during a PhD or post-doc. It requires a willingness to embrace uncertainty and often involves handling diverse tasks without extensive support networks.

ADVANTAGES OF WORKING IN A LARGE CHARITY

- Altruism over profit
- Structured yet adaptable
- Diverse network and relationships
- Clear mission with varied applications
- Broad stakeholder engagement
- More opportunities for part-time and flexible hours

DISADVANTAGES OF WORKING IN A LARGE CHARITY

- Complex organisational dynamics
- Unending commitment
- Versatility required
- Operational responsibilities
- Accountability to a wider Board
- Comparatively lower pay

FILL THE GAPS IN YOUR CV ON YOUR OWN

- Stay informed on relevant issues and policies
- Engage with updates from various charities
- Volunteer - also considering remote opportunities

FILL THE GAPS IN YOUR CV IN YOUR WORK

- Participate in strategy-focused committees
- Collaborate with charities relevant to your research area.
- Engage in academic competitions and presentations
- Document and report your academic progress and goals
- Enhance skills in data management (e.g., Excel)
- Be active on professional networking platforms
- Contribute to newsletters and social media
- Organise meetings with non-academic partners
- Attend and present at external conferences
- Manage volunteers and students
- Actively seek and apply for grants

GREEN FLAGS IN JOB DESCRIPTIONS

- Transparency about staff members and their roles
- Public recognition of staff achievements
- Clear salary details and benefits in job advertisements
- Generous holidays and pension contributions
- Support for remote working and flexible hours

RED FLAGS IN JOB DESCRIPTIONS

- Focus only on board members
- Vague or absent salary details
- Descriptions suggesting high-pressure environments
- Requirements for sudden travel and extended hours
- Emphasis on an involved board and a need for flexibility

THE BEST COURSES TO TAKE

Certificate in non-profit essentials

`nonprofitready.org/free-nonprofit-certificate`

This program offered by the NonprofitReady organisation encompasses three main courses: understanding the structure and roles within non-profits, learning how non-profits raise funds and report on fundraising, as well as guidance on building a successful career in the non-profit sector. Upon completing these courses and passing the certificate exam, participants can earn a certificate and badge that they can display on their CV.

Working in the voluntary sector

`open.edu/openlearn/society-politics-law/sociology/working-the-voluntary-sector/content-section-overview`

This is a free online program that is designed to introduce people to the world of volunteering and working in the voluntary sector. This eight-week course covers various key activities undertaken by volunteers and staff in different organisations, providing a broad understanding of how these entities operate. The course not only offers practical knowledge and skills applicable to work or volunteering but also provides career development in working with volunteers. Participants have the opportunity to earn an Open University digital badge upon completion that they can display on their CV.

~

SAMPLE COVER LETTER

Dear Ms Reibey,

I am writing to express my interest in the Research Coordinator position at the Bright Future Trust, as advertised on CharityJobs.co.uk. With a recently completed Master's degree in Life Sciences from the University of Meyruelle, I am eager to apply my academic background and passion for humanitarian work to contribute meaningfully to your team.

My final thesis project has honed my skills in research, data collection, and project management. I volunteer with Green Steps Meyruelle, where I gained experience in team collaboration, communicating with diverse groups, and organising community engagement events. This experience reinforced my commitment to environmental conservation and equipped me with practical skills in public outreach and project coordination.

I am particularly drawn to the Bright Future Trust because of its commitment to sustainable environmental solutions and community involvement. I am excited about the opportunity to contribute to projects focused on urban conservation and public awareness, and I am confident that my background in life sciences, combined with my passion for environmental conservation, will enable me to make a significant contribution to your team.

Thank you for considering my application. I look forward to the opportunity to discuss how my skills and enthusiasm align with the needs of your organisation. Please find my resume attached for more detailed information about my academic and volunteer experiences.

Yours sincerely,

Eleanor Smythe

~

SAMPLE CV

Eleanor Smythe

A motivated graduate from the University of Meyruelle, specialising in Conservation Biology. With a strong foundation in scientific research and data analysis, I bring a proven track record of leading and participating in environmental conservation projects. Dedicated to utilising my skills in a dynamic, real-world setting to make a tangible impact on conservation efforts.

EXPERIENCE

Volunteer Coordinator

Green Steps Meyruelle, Meyruelle

July 2092 - Present

- Successfully led a team of 30 volunteers in local environmental projects
- Organised 12 community engagement events annually, focusing on urban conservation, with an average attendance of 52 residents per event
- Developed 5 educational outreach programs, enhancing community awareness on environmental conservation

Research Assistant (Part-Time)

University of Meyruelle, Meyruelle

October 2091 - June 2093

- Participated in a two-year biodiversity research project, contributing to 3 peer-reviewed publications.
- Conducted over 200 hours of fieldwork and data analysis, using SPSS and GIS tools.
- Presented research findings at 3 national scientific conferences.

Community Garden Volunteer

Cliffport City Community Gardens

September 2098 - August 2090

- Contributed to the development of 3 community gardens, each spanning over 1000 square feet
- Facilitated 15+ gardening workshops, engaging with over 200 local residents
- Managed the creation and distribution of the annual report

EDUCATION

MSc in Life Sciences

University of Meyruelle, Meyruelle

September 2090 - June 2093

- Dissertation: "Urban Wildlife Adaptation Strategies in Northern England"
- Advanced Statistical Analysis, Ecological Modelling
- Chair of the University of Meyruelle Student Society for Social Action

BSc in Biology

University of Cliffport, Cliffport

September 2097 - June 2090

- Graduated with First-Class Honours
- Final Year Project: "Impact of Climate Change on UK Coastal Ecosystems," achieving a 95% grade
- President of the University of Cliffport Student Society for Diversity and Inclusion

CERTIFICATES

- Futures and foresight in non-profit organisations, *Humanitarian Leadership Academy*
- Safeguarding essentials for charity workers, *Humanitarian Leadership Academy*
- Non-profit essentials, *Non Profit Ready*
- Working in the voluntary sector, *Open University*

SKILLS

- Proficient in SPSS, R, and GIS tools, with over 500 hours of practical application.
- Demonstrated ability in leading projects and coordinating teams effectively.
- Skilled at presenting complex scientific data, having delivered 13 presentations at academic and community events.
- Proven teamwork skills, successfully leading diverse groups in academic and volunteer settings.

PUBLICATIONS

- Smythe, E. (2093). "Urban Green Spaces: Enhancing Biodiversity in Metropolitan Areas." *Journal of Urban Ecology.*
- Smythe, E., & Jones, A. (2092). "Community-Led Conservation: A Model for Sustainable Environmental Practices." *Environmental Sustainability Reports.*
- Smythe, E. (2091). "The Impact of Climate Change on UK Coastal Ecosystems: A Predictive Analysis." *Climate Research Journal.*

HOBBIES

- Urban sketching, wildlife photography, cycling tours

SUMMARY

- **Working in a large charity combines passion and professionalism, offering diverse roles from policy advocacy to stakeholder collaboration, with significant societal impact.**
- **Skills involve essential technical competencies in Customer Relationship Management software, data analysis, and digital marketing.**
- **This sector can demanding adaptability and a wide skill set while providing altruism, adaptable work environments, and diverse networking opportunities.**

CHAPTER 5
MUSEUM

A visit to a museum is a search for beauty, truth, and meaning in our lives.

MAIRA KALMAN

Working in a museum offers a unique and enriching experience, blending a passion for history, art, or science with a professional environment. In a museum, you contribute to cultural and educational enrichment while engaging with a wide range of visitors and collaborators. The impact of your work in a museum is tangible, as you help preserve and present important artefacts and knowledge.

It offers the unique and exhilarating opportunity to engage closely with fascinating artefacts, ranging from ancient dinosaur bones to priceless manuscripts. Each day in a museum can be a journey through time, offering a hands-on experience with items that most people only get to see in documentaries. The privilege of being part of a team that brings these incredible pieces of history to life for the public is both a professional achievement and a personal thrill.

Moreover, the joy of taking friends and family through an exhibit that you've organised or showing them an artefact that you've researched and helped to display is immensely gratifying. It's a chance to share a piece of your professional world with those you love.

In a museum, you often work within a larger team, similar to larger organisations. This can mean navigating through layers of administration, but it also provides access to specialised roles and resources. Your work is driven by the museum's mission to educate and inspire, leading to meaningful experiences for visitors.

Travel may be part of the job for research, attending conferences, or collaborating on exhibitions. Decision-making can sometimes be rapid or may require careful consideration within the organisational structure.

Technical competence is key in the museum sector. This includes expertise in Collection Management Systems for artefact handling, data analysis tools for visitor research, digital marketing for online engagement, Content Management Systems for maintaining the museum's website, and project management tools for organising exhibitions and events. Skills in financial management, online engagement, cybersecurity, cloud services, and social media are also important, supporting the museum in various operational aspects.

Challenges in the museum sector include bureaucratic complexities due to the scale and scope of operations, which can slow down decision-making processes and create administrative hurdles. The nature of museum work can be demanding, driven by the ongoing need to preserve and interpret collections. This requires a high level of commitment and versatility, as well as a broad skill set to meet the diverse demands of museum operations. Salaries in museums are often lower than in some private sector jobs, reflecting the cultural and educational focus of the work.

ADVANTAGES OF WORKING IN A MUSEUM

- Contributing to the preservation and dissemination of knowledge
- Public Engagement
- Diverse and enriching work environment
- Engagement with history, art, culture or science
- Opportunities for research and travel
- Flexible hours and part-time opportunities

DISADVANTAGES OF WORKING IN A MUSEUM

- Complex organisational structure
- Jobs can be highly sought after and competitive
- The workload can be intense
- High commitment and adaptability required
- Limited career advancement opportunities
- Operational and administrative responsibilities
- Comparatively modest pay

FILL THE GAPS IN YOUR CV ON YOUR OWN

- Stay informed about developments in your field
- Volunteer at your local museum
- Engage with museum updates and exhibitions

FILL THE GAPS IN YOUR CV IN YOUR WORK

- Participate in committees focused on exhibition planning or educational outreach

- Collaborate with museums
- Engage in research relevant to museum collections
- Write regular reports on your research progress
- Enhance skills in data management and curation
- Be active on professional networking platforms
- Contribute to museum newsletters or social media
- Organise museum events and public talks
- Attend and present at museum and cultural conferences
- Gain experience in leading volunteer teams

GREEN FLAGS IN JOB DESCRIPTIONS

- Transparency about staff roles on the website
- Public recognition of staff achievements
- Clear salary details and benefits
- Support for work-life balance and flexible hours

RED FLAGS IN JOB DESCRIPTIONS

- Lack of transparency about the full team
- Vague salary details
- Indications of high-pressure work environments
- Requirements for sudden travel

THE BEST COURSES TO TAKE

Culture, health and wellbeing

www.ucl.ac.uk/short-courses/search-courses/
culture-health-and-wellbeing-introduction

This 9 hour free course helps develop, deliver, and evaluate health and wellbeing work within museums, arts, heritage, and cultural organisations. It focuses on partnership working, co-creation with audiences, measuring impact, and safeguarding and influencing senior staff.

The Museum as a site and source for learning

`futurelearn.com/courses/museum-learning`

This course focuses on how museum collections can enhance our understanding of history and how museums can use artefacts to engage learners. It also looks at how museums can better use their spaces to create learning experiences for all visitors. It teaches how to create a museum exhibition that is dynamic and responsive to the needs of museum visitors.

Contemporary museum education

`kadenze.com/courses/contemporary-museum-education/info`

This course offers a blend of theory and practice in contemporary museum education. It examines the shift in art and design education within museums, the changing role of museums as educational entities, and unique learning and interpretation theories in the museum context. There's a practical focus on techniques for teaching with artworks and creating educational materials and programs in art and design museum settings. Additionally, the course delves into critical issues in the field, such as audience diversity, collaboration with educational institutions and communities, reevaluating museum missions, and the potential for innovative practices.

~

L. E. MAKAROFF

SAMPLE COVER LETTER

Dear Dr Jones,

I am writing to express my keen interest in the Collections Care Coordinator role at The Museum of Everything, as advertised on museumjobs.com. With a Master's degree in Mythology and Folklore from St Luke's University, Meyruelle, and extensive experience in European folklore research, I am excited about the opportunity to contribute to your museum.

During my tenure as a Research Assistant at the Meyruelle Folklore Museum, I developed an understanding of the cultural and historical significance of folklore across Europe. My work involved conducting research, curating exhibitions, and engaging the public through educational programmes.

What excites me about this opportunity is the chance to work in an environment that values creativity and innovation. I am impressed by your recent exhibitions, which demonstrate a commitment to historical accuracy and engaging storytelling.

I have a proven track record of working effectively both independently and as part of a team. My ability to communicate complex ideas in an accessible and engaging manner has been a key component of my success in previous roles, and I am eager to bring this skill to The Museum of Everything. I am also excited about the prospect of contributing to a team that is as passionate about education and cultural preservation as I am.

I look forward to the possibility of discussing this exciting opportunity with you. Thank you for considering my application. I am enthusiastic about the prospect of contributing to the continued success of The Museum of Everything.

Yours sincerely,

Abraham Van Helsing

54

~

SAMPLE CV

Abraham Van Helsing, MSc

A dedicated and knowledgeable professional with a Master's degree in Mythology and Folklore from St Luke's University. Specialising in European folklore, with a focus on the socio-cultural impact of mythological narratives. Eager to bring my expertise in research, curation, and public engagement.

EXPERIENCE

Research Volunteer

Meyruelle Folklore Museum, Meyruelle, UK

July 2098 - Present

- Conducted in-depth research on over 20 European folk tales and mythological entities for new exhibits.
- Collaborated on the "Myths of Medieval Europe" exhibition, which attracted 30,000+ visitors.
- Led 5 public engagement sessions, including talks and interactive workshops.

Project Intern

The Historical Society, Manchester, UK

June 2095 - August 2096

- Assisted in cataloguing 1,000+ ancient manuscripts and artefacts.

- Played a key role in organising 15 historical seminars and events.
- Contributed to the development of an interactive digital archive, increasing visitor engagement by 25%.

Volunteer Guide

Manchester Museum, Manchester, UK

September 2093 - May 2095

- Conducted 40+ guided tours focusing on mediaeval European history.
- Assisted in the organisation of 10+ special events and exhibitions.
- Developed a volunteer-led initiative to create multimedia tour content

EDUCATION

MSc in Mythology and Folklore

St Luke's University, Meyruelle, UK

September 2096 - June 2098

- Dissertation: "Transylvanian Folklore and Its Impact on Modern European Culture" (Scored 85%)
- Completed a field study on 5 historic Transylvanian sites, linking folklore to local tourism strategies
- Engaged as a peer reviewer for the university's journal

BA in History

University of Manchester, UK

September 2092 - June 2095

- Specialisation in European Medieval History (Top 5% of class)
- Organised a student-led symposium on medieval European conflicts, attended by 100+ participants
- Contributed to the university's blog, writing 21 articles

CERTIFICATES

- Contemporary Museum Education
- The Museum as a Site and Source for Learning
- Culture, Health and Wellbeing in Museums

SKILLS

- In-depth knowledge of European folklore and mythology.
- Experienced in research and analysis of historical texts and artefacts.
- Proficient in public speaking; delivered 30+ public lectures and workshops.
- Skilled in the use of digital cataloguing systems and Microsoft Office Suite.
- Language Proficiency: Fluent in English, conversant in Romanian.

INTERESTS

Amateur fencing, blogging about mythology

~

L. E. MAKAROFF

SUMMARY

- **Working in a museum offers a fulfilling opportunity to make a difference in cultural preservation and education.**
- **It provides a stimulating environment with varied responsibilities and the chance to engage deeply with your field of interest.**
- **It requires adaptability, a commitment to cultural values, and readiness to handle diverse operational tasks.**

CHAPTER 6
CIVIL SERVICE

*Defining myself, as opposed to being defined by others,
is one of the most difficult challenges I face.*

AUDRE LORDE

The civil service means different things to different people. Some think of it simply as a group of people who work for the government. Others think of it as a group of professionals who work in government agencies to provide services to the public.

The civil service includes all government employees who are not members of the military or the police force. This includes all international, national, state, and local government employees - working for the local council up to the European Commission.

The civil service is responsible for the day-to-day operations of the government and government organisations. They carry out the policies and procedures the elected officials have implemented. The civil service is made up of a variety of different occupations, including clerical workers, lawyers, accountants, and engineers.

The civil service is an integral part of the government because it provides the services it needs to function. Without the civil service, the government would not be able to offer the same level of service to the public.

One significant advantage of a career in the civil service is the supportive work environment regarding work-life balance. Civil service positions often offer more flexibility than many private sector roles. This flexibility includes easier access to parental leave and carer's leave, which is particularly beneficial for employees with young families or caregiving responsibilities. Additionally, the option to move to part-time employment is more readily available, allowing employees to adjust their work commitments to better suit their personal circumstances.

Certain challenges can come with a career in the civil service. One such aspect is the potential need to move between various departments. This could be a source of instability for you, requiring adaptability and a willingness to constantly engage with new teams and environments.

Another challenge is the possibility of having to work for a government that does not align with your personal political views. Civil servants are required to execute policies impartially, regardless of their own beliefs or the party they voted for. This can sometimes create internal conflicts, especially when working on policies that one might not personally support. You may also have to abide by a policy that states you will not be able to express your opinions about the government on your personal social media.

Moreover, certain types of civil service jobs, particularly those in specialised fields, may only be available in capital cities. This geographic limitation can be a barrier for those who prefer or need to live in other regions, potentially leading to difficult decisions about relocation.

Red tape is another factor that can be frustrating in the civil service. The need for thorough procedures and adherence to strict regulations can slow down decision-making processes, which can be challenging for those who prefer a more dynamic and fast-paced work environment.

Finally, budget cuts can significantly impact the ability to deliver services effectively. These cuts can lead to resource constraints, making it challenging to meet the needs of the public efficiently. This aspect of the job can be particularly disheartening, as it will directly affect your ability to contribute positively to society.

ADVANTAGES OF WORKING IN THE CIVIL SERVICE

- Wide range of roles across various departments
- Comprehensive training and development
- Great retirement package
- Great work/life balance
- Job security
- Opportunities for professional development
- The chance to make a positive impact on society

DISADVANTAGES OF WORKING IN THE CIVIL SERVICE

- You might have to move around various departments
- You might have to work for a government that you didn't vote for
- Some civil service jobs are only available in capital cities
- Red tape can slow down decision-making
- Budget cuts can impact the ability to deliver services

FILL THE GAPS IN YOUR CV ON YOUR OWN

- Learn more about your government institutions
- Learn about health policy
- Volunteer in something that lets you manage people
- Write a monthly report to your supervisor
- Learn another language

FILL THE GAPS IN YOUR CV IN YOUR WORK

- Enter competitions for posters and presentations
- Take a course in presentation skills
- Take a class in Excel skills
- Ask to organise a meeting that involves external people
- Travel to external congresses and give oral presentations
- Ask to manage volunteers and students

GREEN FLAGS IN JOB DESCRIPTIONS

- Clear responsibilities and expectations outlined
- Opportunities for growth and advancement
- Salary is mentioned
- Generous holiday allowance
- Additional pension contributions are stated
- Remote working or hybrid working is encouraged
- Flexible working hours

RED FLAGS IN JOB DESCRIPTIONS

- Salary is not mentioned
- Vague job descriptions
- Qualifications listed for the job seem overly demanding
- The job is offering very low pay or limited benefits

- The job description doesn't mention opportunities for career growth or professional development
- "ability to work under pressure"
- "vacation to be taken according to the needs of the office"
- "ensure all projects operate flawlessly"

THE BEST COURSE TO TAKE

Public Financial Management

`www.edx.org/course/public-financial-management`

What is the government budget cycle? How should governments prepare policy-oriented budgets? How to hold governments accountable? Learn from an International Monetary Fund team that advises on budget management and hear testimonies from ministers of finance and civil society.

∽

SAMPLE COVER LETTER

Dear Ms Rodríguez,

I am writing in response to your advertisment for a Programme Specialist published on your website. The most effective evidence-based policies must be based on accessible data, so I would value the opportunity to be involved with this department.

I am an international scientist with experience in generating and communicating data. I am currently working as a post-doctoral research fellow at the Unseen University, and I hold a PhD in immunology and a Bachelor of Psychology in advanced analytical methodology. I have experience in dataset management and analysis. During my academic career, I have designed and managed multiple projects while supervising junior scholars. I have published my studies in high-impact journals and presented these results at international conferences. I have been recognised with awards for my outstanding communication abilities. I am capable of productive multidisciplinary collaboration with other researchers.

As a computer consultant for Banting University, I used my information technology skills to train staff and students in many areas of computing, including Microsoft Office applications. I also have extensive experience in the use of many statistical packages, such as R, SAS, SPSS, S-Plus, and Prism. I am a native English speaker with excellent communication skills, and I am currently learning French. I am motivated, enthusiastic, and dedicated to achieving excellence in classification and communication.

Furthermore, my ability to adapt to diverse work environments and my experience in managing cross-functional teams will be an asset in this role. My goal is to apply my comprehensive skill set to effectively contribute to your department's objectives and support the development of impactful policies. I am excited about the prospect of joining your team and bringing a fresh perspective, backed by a strong foundation in scientific research and data analysis.

I can be reached via email at Tang@u.unseen.edu or via phone on +1 206 240 6291. I hope to hear from you about this employment opportunity.

Sincerely,

Sam Tang.

~

SAMPLE CV

Dr Sam TANG, PhD

Seeking a position in the European Commission to apply my analytical and project management skills towards the advancement of the European Union.

EXPERIENCE

Post-doctoral research fellow

Unseen University, 2099-present

- Conducted research and analysis on cancer
- Designed and managed research projects
- Produced reports and publications for academic and non-academic audiences
- Developed policy recommendations
- Collaborated with stakeholders to implement policy changes

EDUCATION

PhD in Medical Research, *Unseen University*, 2099

Bachelor of Science in Psychology, *Banting University*, 2095

- Published 2 peer-reviewed articles in high-impact journals, cited over 20 times
- Received the 'Shining Star' award for outstanding research presentation
- Acted as a peer reviewer for 3 journals

Skills

- Certificate in Public Financial Management
- Expert in statistical software and data management tools
- Mentorship experience with junior researchers
- Strong collaboration and partnership-building abilities

Languages

- Native proficiency in English
- Professional proficiency in French
- Basic proficiency in German

Hobbies

Robotics, orienteering, vexillology

∾

SUMMARY

- **The civil service encompasses government employees who provide services to the public and carry out government policies and procedures.**
- **It offers benefits such as job security, work/life balance, retirement packages, and opportunities for professional development.**
- **Challenges include potential department transfers, working under a government one did not vote for, limited job availability outside capital cities, bureaucratic red tape, and budget cuts impacting service delivery.**

CHAPTER 7
HEALTH AND SAFETY OFFICER

Anyone can hide. Facing up to things, working through them, that's what makes you strong.

SARAH DESSEN

W as the favourite part of your PhD or post-doc organising the lab? Did you take reading the safety guidance more seriously than those around you? Do you enjoy problem solving, attention to detail, and have good people skills? Do you feel comfortable learning about legal and regulatory matters? If so, then a job as a Health and Safety Officer might be for you.

A Health and Safety Officer in a research institute works to ensure that the institute is complying with all health and safety regulations. They conduct inspections of the facilities and equipment to identify potential hazards and work with the institute's staff to develop and implement health and safety protocols. The officer also provides training on health and safety topics, and they may investigate accidents or incidents at the institute.

A typical day in the life of a Health and Safety Officer is diverse and involves several key responsibilities aimed at maintaining a safe work environment. The day often begins with conducting safety audits, a detailed process where the officer inspects various aspects of the workplace. This includes checking machinery, equipment, and facilities to ensure they meet safety standards and identifying potential hazards.

Developing and implementing safety initiatives is a proactive aspect of the role. This involves creating policies and procedures that promote a safe working environment. The officer might launch campaigns focused on specific safety issues, such as proper handling of hazardous materials or ergonomics in the workplace.

Developing safety training programs is vital for educating employees about workplace hazards and safe practices. The Health and Safety Officer designs these programs, tailors them to various departments, and often delivers the training sessions themselves. This training ensures that all employees are aware of safety protocols and know how to respond in emergency situations.

Collaboration with other departments is also a significant part of the role. The Health and Safety Officer works closely with department heads and managers to ensure that safety procedures are integrated into all work processes. They also ensure that safety equipment, like fire extinguishers, safety goggles, and protective clothing, is not only available but also properly used by staff.

In addition to these tasks, Health and Safety Officers are often involved in regular meetings with senior management to report on safety issues, provide updates on initiatives, and advise on compliance with health and safety legislation. They stay abreast of legal and regulatory changes in their field and revise company policies accordingly.

The best part of working as a Health and Safety Officer is that you help ensure everyone stays safe while working. You also get to work with a great team of people who are all committed to keeping your workplace safe. You can use your scientific knowledge and problem-solving skills, and get to be involved in lots of projects without being based at the bench.

The worst part of working as a Health and Safety Officer is that you are always on the lookout for potential hazards. It's a job with mainly negative performance indicators. You'll never know how many accidents didn't happen, only the ones that did. Additionally, Health and Safety Officers often must deal with difficult people who may resist change or do not take safety seriously.

ADVANTAGES OF BEING A HEALTH & SAFETY OFFICER

- You can still work in a research institute
- Good work/life balance
- High job security
- Able to move quickly to another institute
- Ability to move up into management roles if desired
- You are enabling cool, exciting things to happen

DISADVANTAGES OF BEING A HEALTH & SAFETY OFFICER

- Your name isn't going to be on any more papers
- Your job has serious responsibilities
- Employees may resist changes as you encourage a culture of safety
- In social settings, there might be a preconceived notion you're overly strict

- Substantial amount of paperwork and administration

FILL THE GAPS IN YOUR CV ON YOUR OWN

- Join a local union
- Join a local national professional body, e.g. IOSH in the UK (www.iosh.com) which anyone can join as an affiliate member

FILL THE GAPS IN YOUR CV IN YOUR WORK

- Volunteer to work on a safety committee
- Volunteer to review and update your lab's safety sheets
- Volunteer to give presentations on lab safety
- Ask to participate in any risk assessments
- Talk to your institute's Health & Safety team and ask if there are ways you can get experience or join in with any inspections
- Volunteer for named safety roles such as Fire Marshal or First Aider

GREEN FLAGS IN JOB DESCRIPTIONS

- Emphasis on workplace safety
- Mentions a collaborative work environment with opportunities for teamwork
- Clear responsibilities and expectations outlined
- Opportunities for growth and advancement mentioned
- Emphasis on continued training and development
- Salary is mentioned in the job advertisement
- Generous holiday allowance

RED FLAGS IN JOB DESCRIPTIONS

- Salary is not mentioned
- Vague job descriptions
- The qualifications listed for the job seem unrealistic
- Includes high number of duties
- Does not include any information on training
- The job is offering very low pay or limited benefits
- "ability to work under pressure"
- "young team"
- "ensure all projects operate flawlessly"
- "availability to undertake travel at short notice"
- "willingness to work longer hours when needed"
- "character traits of being flexible and stress-resistant"

THE BEST COURSES TO TAKE

The best course to take is a professional qualification — one about health & safety as a discipline, rather than how to work safely. In the UK, the National Examination Board in Occupational Safety and Health (NEBOSH) offers a national general certificate in occupational health and safety, highly regarded in the industry. In the USA, the Board of Certified Safety Professionals provides certifications like the Certified Safety Professional. The Canadian Registered Safety Professional certification, awarded by the Board of Canadian Registered Safety Professionals, enjoys widespread recognition in Canada. In Australia, the practical and widely pursued Certificate IV in Work Health and Safety prepares individuals for roles in this field. For New Zealand, the New Zealand Certificate in Workplace Health and Safety Practice, available at Levels 3 and 4, lays a solid foundation for health and safety practices in the local work environment. Each of these qualifications is designed to ensure Health and Safety Officers are thoroughly prepared with the knowledge and skills pertinent to the specific legal and regulatory frameworks of their respective countries.

If you're not ready to sign up for a paid professional health & safety qualification, you may wish to take:

Health and safety in the laboratory and field

```
www.open.edu/openlearn/science-maths-
technology/health-and-safety-the-laboratory-
and-field/content-section-0?active-tab=
description-tab
```

Health, safety, and risk assessment are paramount in the laboratory and the field. This free course will help make you more aware of the hazards and risks involved wherever you undertake your research and give you an overview of the legal requirements attached to this work. The course discusses issues in handling chemical and biological agents, basic safety procedures, and common field-work hazards.

∼

SAMPLE COVER LETTER

Dear Mr Dredd,

I am writing to express my strong interest in the Health & Safety Officer position at the Laboratory of Scientific Molecules, as advertised. With a PhD in Biology from the University of Atlantis and extensive experience in laboratory environments, I am excited about the opportunity to contribute to your world-class research facility.

During my academic career, I have been deeply involved in ensuring safe laboratory practices and compliance with health and safety regulations. My experience includes conducting detailed safety audits, developing safety protocols, and leading training sessions for new staff members. I have honed my skills in identifying and mitigating risks in high-stakes research settings, which aligns perfectly with the responsibilities outlined for this role. My proactive approach in evaluating and improving safety procedures has been instrumental in maintaining a safe working environment.

I am particularly drawn to the Laboratory of Scientific Molecules because of its reputation for cutting-edge research and its collaborative culture. The prospect of working in your new state-of-the-art facility within the dynamic Genovia Hospital Campus is particularly appealing. I am confident that my background in biology and safety management will allow me to effectively support the laboratory's diverse research projects while ensuring the highest standards of workplace safety.

As an advocate for continuous learning and improvement, I am keen to bring my expertise to the Laboratory of Scientific Molecules and collaborate with the team to further enhance safety practices. I am enthusiastic about the possibility of joining such an innovative and forward-thinking organisation and contributing to its success.

Thank you for considering my application. I look forward to the opportunity to discuss how my skills and experiences align with the needs of your team. Please find attached my CV and contact details of two scientific references.

Yours sincerely,

Dr Minerva Campbell, PhD

~

SAMPLE CV

Dr Minerva Campbell, PhD

Expert in developing and implementing safety protocols in academic and research environments. Skilled in developing and implementing comprehensive safety protocols, particularly in academic and research settings. Demonstrates exceptional leadership in promoting workplace safety, regulatory compliance, and a culture of continuous improvement in health and safety practices.

EXPERIENCE

PhD in Biology

University of Atlantis, Atlantis

September 2085 - July 2089

- Implemented 15 safety policies on university committee
- Updated 52 safety data sheets to latest regulations
- Delivered 23 safety awareness presentations
- Performed 34 risk assessments
- Participated in 25 safety inspections

Bachelor of Science in Environmental Health

Atlantis University, Atlantis

September 2081 - June 2085

- Graduated with First-Class Honours, top 5% of class
- Led Student Environmental Health Society as President
- Organised 15 workshops, educating 150 students

Certifications and Training

- National Examination Board in Occupational Safety and Health General Certificate in Occupational Health and Safety
- Certified in Fire Safety Management Training
- Advanced First Aid, Cardiopulmonary Resuscitation, Automated External Defibrillator Certification

Memberships

- Involved the Institution of Occupational Safety and Health, serving as an affiliate member
- Contributor to the Safety and Health Practitioners online community
- Member of the International Institute of Risk and Safety Management

Skills

- Expertise in hazardous material handling
- Safety auditing and risk assessment techniques
- Strong training and leadership capabilities
- Excellent knowledge of health and safety legislation
- Effective communication and interpersonal skills
- Fluent in English and conversant in French

～

SUMMARY

- A career as a Health and Safety Officer involves ensuring compliance with regulations, conducting safety audits, developing protocols, and providing training.
- The advantages of the role include the opportunity to work in a research institute, a good work/life balance, job security, and the ability to contribute to exciting projects.
- Disadvantages include not having your name on research papers, significant responsibilities, and occasional social challenges in discussing the job.

CHAPTER 8
TECHNICIAN

It's not happening as fast as you'd like, but it is happening.

RAMIN NAZER

Was working at the bench your favourite part of your PhD or post-doc? Did you find getting experiments easier than those around you? If so, then a job as a technician might be for you.

Technicians play an indispensable role in research institutes, hospitals, or pharmacies. Their responsibilities extend beyond basic maintenance; they are tasked with operating sophisticated equipment and instruments, executing experiments, gathering data, and efficiently managing inventory.

A day in the life of a laboratory or hospital technician is characterised by its variety and dynamism, offering a blend of technical skills and critical thinking. The day typically begins with the processing of samples, a task that demands precise attention to detail and accuracy. This involves carefully preparing and handling biological or chemical specimens, ensuring that they are correctly labelled, stored, and processed for analysis. The accuracy in this initial stage is fundamental, as it sets the foundation for reliable test results and research outcomes.

Maintaining laboratory equipment is another vital aspect of a technician's role. This not only includes routine cleaning and calibration of instruments but also troubleshooting and minor repairs. Ensuring that all equipment is in optimal working condition is crucial, as any malfunction can lead to erroneous results or delay important research activities. A technician's ability to quickly diagnose and resolve equipment issues is essential for the smooth operation of the laboratory.

Record-keeping is another critical component of a technician's responsibilities. This involves diligently documenting experimental procedures, results, and any observations. Good record-keeping practices are essential for maintaining the integrity of the research process. They enable other researchers to replicate experiments, verify results, and build upon previous work. This meticulous approach to documentation also extends to maintaining logs for equipment usage, servicing, and repairs, which are crucial for quality control and regulatory compliance.

Technicians often work closely with researchers, clinicians, and other healthcare professionals, providing technical support and expertise. They may be involved in setting up experiments, preparing reagents and solutions, and ensuring that all necessary materials are available and in good condition. Their role is pivotal in facilitating the research process, from experimental design to data collection and analysis.

The dynamic nature of a technician's job means that each day brings new challenges and learning opportunities. Whether it's adapting to new protocols, managing urgent sample processing requests, or staying updated with the latest techniques and technologies, technicians are constantly evolving and expanding their skill set. This continuous development is not only professionally rewarding but also ensures that they remain valuable and effective members of the laboratory or hospital team.

Sometimes your experiment won't work, and no one knows why. Sometimes it won't work even though it worked before. You might spend days or weeks trying to determine what happened and why. You will also spend your entire life setting timers, watching timers, and fighting the urge to smash them when they go off.

Being a technician is a role that offers a blend of practical skills, problem-solving, and the opportunity to contribute significantly to the advancement of scientific research. For those who thrive in a hands-on, dynamic environment, and possess a keen eye for detail, it's a career path worth considering.

ADVANTAGES OF BEING A TECHNICIAN

- Ability to work part-time
- You can still work in a laboratory
- You can leave the work at work
- Opportunity to learn new techniques and skills
- You can gain skills that can be applied to pharmaceuticals and biotechnology

DISADVANTAGES OF BEING A TECHNICIAN

- Fewer opportunities for career progression
- Mistakes can cost your employer thousands of dollars
- You may be working with dangerous chemicals
- Harder to work from home
- Fewer opportunities to learn transferable skills

FILL THE GAPS IN YOUR CV ON YOUR OWN

- Take online training courses from LabCE and LabeXchange
- Build relationships by joining LabWrench and LabRoots online forums
- Develop skills in data analysis and coding

FILL THE GAPS IN YOUR CV IN YOUR WORK

- Keep an eye out for any technical training courses offered by your university
- Talk to people in other labs and ask if you can learn their techniques too
- Volunteer to write up any techniques or protocols
- Keep up with advances in laboratory techniques by attending workshops
- Build professional relationships with others in the field by attending events
- Study the laboratory safety protocols

GREEN FLAGS IN JOB DESCRIPTIONS

- Emphasis on workplace safety
- Mentions a collaborative work environment
- Clear responsibilities and expectations outlined
- Opportunities for growth and advancement
- Emphasis on continued training and development
- Salary is mentioned in the job advertisement
- Generous holiday allowance
- Additional pension contributions are stated
- Remote working or hybrid working is encouraged
- Flexible working hours

RED FLAGS IN JOB DESCRIPTIONS

- Requires being on call 24/7
- Salary is not mentioned
- Vague job descriptions
- Qualifications listed for the job seem overly demanding
- Includes an high number of duties
- No information on training opportunities
- The job is offering very low pay or limited benefits
- "ability to work under pressure"
- "ensure all projects operate flawlessly"

THE BEST COURSES TO TAKE

Repairing and Maintaining Biomedical Devices

www.edx.org/course/biomedical-equipment-
technician-training-maintenance-repair

Understand the workings of biomedical devices. Completing this course adds a complementary skill set to the prior competence of the learner to perform maintenance routines and diagnose problems in sophisticated equipment with care. On a personal level, the learner can rely on the knowledge gained from this course as 'the blueprint' for whenever they have to maintain/troubleshoot any biomedical device.

MATLAB Programming for Engineers and Scientists

coursera.org/specializations/matlab-
programming-engineers-scientists

This course is designed to transition learners from little to no programming experience to becoming proficient in creating MATLAB programs that solve real-world problems in engineering and science. This course covers a range of topics from basic programming concepts to more advanced techniques including recursion, program efficiency, Object-Oriented Programming, graphical user interfaces, and machine learning. Participants will gain hands-on experience with image processing, data visualisation, and applying machine learning methods for data classification and prediction. This three-course series, which includes two practical projects, aims to provide a deep understanding of MATLAB and its application in scientific computing and data analysis.

White Blood Cell Differential Simulator

www.labce.com/white-blood-cell-wbc-case-simulator.aspx

A resource for practising white blood cell identification. The LabCE White Blood Cell Differential Simulator includes 25 expert-reviewed differentials, each with 100 slide images. Perform the differential yourself and then compare your cell identifications with the experts.

~

SAMPLE COVER LETTER

Dear Mr Grubb,

I am writing to express my interest in the laboratory technician position at Unseen University, advertised on LinkedIn. As a PhD graduate from the University of Arendelle, I have extensive experience in laboratory research. I am excited about the opportunity to apply my skills and knowledge in a professional setting.

During my PhD studies, I conducted various experiments involving both wet and dry lab techniques, including DNA extraction, polymerase chain reactions, gel electrophoresis, protein purification, quantum cell sequencing, nanozyme synthesis, cryogenic electron microscopy, photon flux spectroscopy, and bio-neural gel pack analysis. I also have experience in microscopy, flow cytometry, and data analysis. In addition, I have experience in maintaining and troubleshooting laboratory equipment, ensuring a safe laboratory environment, and managing laboratory inventory.

As a laboratory technician at Unseen University, I would bring my strong attention to detail, problem-solving skills, and ability to work collaboratively with others to the team. I am committed to contributing to the success of the laboratory and am eager to learn new techniques and technologies.

Furthermore, my experience in academic research has equipped me with excellent written and verbal communication skills, crucial for maintaining accurate lab records and presenting findings. I am adept at working under pressure and managing multiple tasks efficiently, ensuring the timely completion of projects without compromising on quality. My proactive approach and dedication to ongoing learning make me a perfect fit for the dynamic and innovative environment at Unseen University.

Thank you for considering my application. I look forward to the opportunity to discuss my qualifications further.

Sincerely,

Anna Hansen

SAMPLE CV

Dr Anna Hansen, PhD

Detail-oriented Biological Sciences PhD graduate, with a comprehensive background in laboratory techniques and safety protocols. My dedication to scientific research is demonstrated through my extensive hands-on experience, innovative problem-solving skills, and commitment to advancing the field of biology.

EXPERIENCE

Laboratory Teaching Assistant

University of Arendelle

- Assisted in laboratory classes for undergraduate students, averaging 30 students per session
- Supervised students in performing over 15 different laboratory techniques and experiments
- Graded and provided feedback on approximately 200 lab reports per semester
- Managed the preparation and maintenance of laboratory equipment and reagents for each class, ensuring smooth class operations.
- Coordinated with the IT department to integrate digital tools in laboratory teaching

EDUCATION

PhD in Biological Sciences

University of Arendelle

- Specialised in Quantum Entanglement Microscopy, amassing over 500 hours of practical experience
- Streamlined laboratory inventory processes, achieving a 30% reduction in equipment downtime
- Mentored 42 students in advanced laboratory skills, fostering their scientific development

Bachelor of Science

University of Arendelle

- Thesis on cellular biology, receiving a distinction
- Utilised R for data analysis of 63 datasets
- Actively participated in scientific workshops

Certificates

- Repairing and Maintaining Biomedical Devices
- MATLAB Programming
- Advanced Data Analytics with Python

Skills

- Presented research findings at 5 international conferences and authored 2 papers in peer-reviewed journals.
- Demonstrated verbal and written communication skills through conducting 23 training workshops and seminars.
- Proven strong problem-solving skills and attention to detail, successfully troubleshooting and resolving over 50 technical issues in laboratory settings.

Publications

Sam E. Tang, Nushi W. Kim, Ani E. Ali, and Mary J. Silva. 2129. The Information Paradox. *Proceedings of the National Academy of Sciences* Eddis, in press.

Sam E. Tang, Elena Smith, and Rita Devi. 2128. The Data Loss Challenge. In S Torres and M Marin (Eds), *Genetic Predisposition to Disease*, pp 39-71. Cliffport: Avan Publishers.

Hobbies

Glassblowing, kite making, astrophotography

~

SUMMARY

- **A job as a technician offers the opportunity to work in a laboratory, be involved in research, and learn new techniques and skills.**
- **Advantages include part-time work options, the ability to leave work at work, learning from experienced researchers, and gaining skills applicable to pharmaceuticals and biotechnology.**
- **Disadvantages may include limited career progression, high stakes as mistakes can be costly, exposure to dangerous chemicals, challenges with remote work, and fewer opportunities for transferable skills.**

CHAPTER 9
MEDICAL WRITER, TECHNICAL WRITER, OR JOURNALIST

The bird who dares to fall is the bird who learns to fly.

UNKNOWN

Was your favourite part of your PhD or post-doc when you finally got to sit down and write that paper? Did you find writing your thesis easier than those around you? Did you enjoy the journal club? If so, then a job as a writer might be for you.

A medical writer works with a scientific or medical team to communicate complex information about products, services, and research in various formats. The writer develops and edits educational and promotional materials for a broad audience. A specialist journal, pharmaceutical company, communications agency, or academic institution may employ you.

A technical writer deals with a broader range of subjects, including technology, engineering, and software, where they create manuals, how-to guides, and documentation that simplify complex technical information for end-users. While both roles require excellent writing and research skills, a medical writer's work is more specialised in the healthcare sector, whereas a technical writer covers a wider spectrum of industries and technical subjects.

A journalist typically works on current events and stories, aiming to inform the public through news articles, reports, and features, and often works under tighter deadlines to deliver timely and engaging content on a wide array of topics. You must understand and explain complex scientific and medical information in plain language. You must be able to work independently and as part of a team.

A typical day as a medical writer or journalist is spent mostly researching, then writing and editing text. Medical writers or journalists spend a lot of time researching information and conducting interviews with experts in the field. They then turn this research into original content or use it to update existing content for clarity and accuracy.

A significant part of the job involves gathering information and understanding complex topics. You will need to have a deep understanding of the subject matter you are reporting on, and this requires a lot of time and effort invested in research and interviews. If you decide to pursue journalism, make sure to research and learn the code of ethics that's unique to the field since it's quite different from academia.

You can gain a competitive edge by developing skills in data visualisation. Being able to create compelling infographics and visual representations of complex data is becoming increasingly important in journalism. Taking courses in data visualisation will help you learn how to communicate complex information in a visual format, making it more accessible to your audience.

Build experience in public speaking and video production. As a journalist, you may be expected to participate in podcasts, videos, and moderating panel debates. Developing skills in public speaking, team meeting facilitation, and being comfortable on camera will be helpful in pursuing this aspect of the job.

The best aspects of working as a medical writer or journalist are the variety of topics to learn and write about and the opportunity to work with different people. If you think you would enjoy the challenge of writing about complex or specialised issues, this could also be a good role for you.

Being a writer also comes with its own set of challenges. One notable downside is the potential for work-related stress due to tight deadlines and high expectations for accuracy and detail. Writers often face the pressure of translating complex medical information into straightforward, concise, and engaging content under stringent timelines, which can be demanding and sometimes overwhelming. The job can involve long periods of solitary work, which may lead to isolation for those who thrive in more collaborative and interactive environments. The need to constantly update your knowledge to stay abreast of the latest medical research and guidelines can also be daunting, requiring continuous learning and adaptation. Furthermore, the highly specialised nature of the work can sometimes limit opportunities for creative expression, as the content must adhere to strict medical and scientific standards. These aspects of the job require high resilience and the ability to manage stress effectively to ensure both personal well-being and professional success.

A career in medical writing or journalism offers a unique opportunity to delve into a variety of topics, collaborate with diverse professionals, and make complex scientific knowledge accessible to a broader audience. For those with a passion for writing and a knack for conveying intricate ideas in a comprehensible manner, this career path offers an exciting and fulfilling way to contribute to the scientific community.

ADVANTAGES OF BEING A WRITER

- Varied topics offer intellectual stimulation
- Opportunity to work with and interview experts
- Flexibility in the work environment and hours
- The potential for remote working
- Can contribute to the public understanding of science

DISADVANTAGES OF BEING A WRITER

- Tight deadlines can lead to high stress
- Need to update knowledge in rapidly evolving fields
- Potential for repetitive tasks and topics over time
- Limited creative freedom in highly regulated fields
- Can involve long hours of solitary research and writing

FILL THE GAPS IN YOUR CV ON YOUR OWN

- Start a professional X (formerly Twitter), Mastodon, or Threads account
- Take a course in scientific writing or journalism skills
- Take a course in presentation skills
- Contribute to blogs or other publications
- Volunteer to write for a non-profit organisation
- Create a website containing a portfolio of your writing
- Write for medium.com, Substack, or Ghost
- Install a touch typing program and practise
- Read New Scientist, Wired, and Scientific American
- Find good broadsheet scientific journalists - like Sarah Boseley, Robin McKie and Anjana Ahuja - and follow their outputs

FILL THE GAPS IN YOUR CV IN YOUR WORK

- Write as many review articles as possible
- Write a short monthly report to your supervisor
- Ask your supervisor if you can review articles from peer-reviewed journals
- Ask to draft press releases related to publications
- Contribute to the newsletters and social media
- Volunteer to work on a communications committee that
- Enter competitions for posters and presentations
- Use reference management software – EndNote, etc. – for everything you write
- Travel to congresses and give presentations

GREEN FLAGS IN JOB DESCRIPTIONS

- Clear and specific job responsibilities
- Mention of a supportive team environment
- Emphasis on work-life balance and remote work
- Details about professional development opportunities
- Transparency about compensation and benefits
- Positive language around creativity and innovation
- Respectful and inclusive workplace culture
- Clear descriptions of the reporting structure
- "guidance and mentorship from senior colleagues"
- "access to training opportunities"
- "a team that makes informed decisions together"

RED FLAGS IN JOB DESCRIPTIONS

- Lack of clarity on reporting structure
- Overly broad or unrealistic job requirements
- Expectations of availability, including weekends

- Unpaid trial periods
- Requests for speculative work without compensation
- "Responsibilities may vary and evolve rapidly"
- "Fast-paced, high-intensity environment"
- "Must be willing to work under tight deadlines"

THE BEST COURSES TO TAKE

Scientific writing

`coursera.org/learn/sciwrite`

This course teaches scientists to become more effective writers using practical examples and exercises. Topics include principles of good writing, tricks for writing faster and with less anxiety, the format of a scientific manuscript, peer review, grant writing, ethical issues in scientific publication, and writing for general audiences.

Introduction to technical writing

`coursera.org/learn/technical-writing-introduction`

This course imparts an understanding of technical writing's nuances, history, and distinctions from other writing forms. The course guides students through creating various technical documents, such as user manuals and documentation, while upholding the highest ethical standards and ensuring accessibility. It covers essential aspects of document design, the use of writing tools, and effective collaboration on platforms like GitHub. Additionally, the course prepares students for career growth in technical writing by teaching them how to build a robust portfolio.

Become a journalist

`coursera.org/specializations/become-a-journalist`

This program equips learners with essential journalistic skills for print, broadcast, and social media platforms, starting from the basics of news reporting to advanced techniques. It delves into the ethical standards and best practices of newsgathering, as well as the impact of journalism on societal issues, preparing students for various career opportunities in the field. The course culminates in a Capstone project where students create a professional-quality news report, demonstrating their acquired skills in journalism.

∾

SAMPLE COVER LETTER

Dear Jack Spang,

I am writing to express my interest in the Medical Writer position at your Healthcare Communications agency in Genovia, as detailed on your website. With my PhD in Molecular Biology from Gondor National University as well as my extensive experience in scientific communication, I am well placed to contribute to your team and develop engaging content for various medical communications materials.

During my doctoral research, I authored and published three articles in high-impact, peer-reviewed journals and presented my findings at eight national and five international conferences. This experience honed my ability to communicate complex scientific information clearly and effectively. Additionally, I led a team of four junior researchers, guiding them in data collection, analysis, and scientific writing. My role also involved drafting and acquiring ethical approvals for research protocols, as well as securing funding from multiple sponsors through successful grant proposals.

My commitment to scientific communication was further developed through a rigorous Graduate Award in Scientific Communication, which resulted in an award for best paper at a university symposium. I have also organised and executed several workshops on scientific presentation skills for my peers, as well as being responsible for nine group discussions on scientific topics.

In addition to my academic and research accomplishments, I have developed a keen understanding of various audiences through my experience in public outreach and science communication initiatives. My ability to distil complex scientific concepts into relatable and engaging narratives has been demonstrated through numerous public lectures, community outreach programs, and collaborations with media outlets. I am particularly drawn to the innovative approach your agency takes in healthcare communications, and I am excited at the prospect of contributing my unique blend of scientific expertise and communication acumen to your esteemed team.

I am confident that my strong background in molecular biology, combined with my proven skills in scientific writing and communication, make me an ideal candidate for the Medical Writer role. I am excited about the opportunity to use my expertise to deliver high-quality, accurate, and engaging content to your clients.

Thank you for considering my application. I am enthusiastic about the prospect of joining your team and contributing to the impactful work at your agency.

Yours sincerely,

Sam Tang

~

SAMPLE CV

Sam Tang, PhD

A Molecular Biologist with a PhD from Gondor National University, with expertise that spans a broad spectrum of scientific and statistical analyses, demonstrated through academic output and awards in science and bioinformatics, underscoring a commitment to advancing scientific knowledge and education.

RESEARCH EXPERIENCE AND EDUCATION

Doctor of Philosophy in Molecular Biology

Gondor National University, 2124-2127

- Authored and published 12 scientific articles in high-impact, peer-reviewed journals
- Presented novel scientific results at national and international conferences
- Led a team of junior researchers, fostering their skills in data collection, analysis, and scientific writing
- Drafted and successfully acquired ethical approvals for 10 research protocols
- Secured funding from 6 different sponsors through grant proposals

Graduate Award in Scientific Communication

Gondor National University, 2124

- Enhanced scientific communication skills, resulting in 3 awards for best paper at university symposiums
- Developed and executed 5 workshops on scientific presentation skills for peers
- Led 12 group discussions on scientific topics

SELECTED PUBLICATIONS

Sam E. Tang, Nushi W. Kim, Ani E. Ali, and Mary J. Silva. 2129. *The Information Paradox*. Proceedings of the National Academy of Sciences Eddis, in press.

Sam E. Tang, Elena Smith, and Rita Devi. 2128. *The Data Loss Challenge*. In S Torres and M Marin (Eds), Genetic Predisposition to Disease, pp 39-71. Cliffport: Avan Publishers.

SELECTED PRESENTATIONS

Sam E. Tang, Ani E. Ali, and Mary J. Silva. 2128. Developing and Delivering Exceptional Workloads. *Gene Expression*. Hot Spring Harbor, Eddis.

Sam E. Tang, Ani E. Ali, and Mary J. Silva. 2128. Managing Student Success. *Joint Research Symposium*. Springfield, Eddis.

LEADERSHIP

President, Banting University Students' Association. 2122. Directed allocation of funding and organised events.

Computer Consultant, Banting University. 2121-2127. Worked with clients to improve their IT skills

Demonstrator and Tutor, Banting University. 2124-2127. Responsible for the teaching of undergraduate biology students

SKILLS

- Word, Excel, PowerPoint, Access, Outlook
- HTML, XML, CSS
- R, SAS, SPSS, S-Plus, Prism

HONOURS and PRIZES

- VBNMC Student Travel Bursary (2125)
- LKJHG Bioinformatics Student Seminar Award (2125)
- POIUY Program for Excellence in Science (2124)

PROFESSIONAL AFFILIATIONS

- Gondor Association for the Advancement of Science
- Genovia Society for Immunology

∾

SUMMARY

- **A job as a writer offers the opportunity to communicate complex scientific or medical information in various formats.**
- **Advantages include the variety of topics to learn and write about and the opportunity to work with different people.**
- **Challenges include keeping up with the rapidly changing field and ensuring accuracy in conveying information.**

∾

CHAPTER 10
PHARMACEUTICAL COMPANY

If I'm an advocate for anything, it's to move. As far as you can, as much as you can. Across the ocean or simply across the river. The extent to which you can walk in someone else's shoes or at least eat their food, it's a plus for everybody. Open your mind, get up off the couch, move.

ANTHONY BOURDAIN

My first day working at a pharmaceutical company was overwhelming. Everyone was speaking a different language. Countries were called markets. Diseases were called indications. PowerPoint presentations were called slide decks. My first week was nothing but online training modules about compliance and pharmacovigilance. Everyone was in meetings all the time.

There are many different job descriptions within the pharmaceutical industry. You can work in advocacy, where you collaborate with patient and professional groups to increase awareness of diseases. You can work as a Medical Science

Liaison, where you work in the field developing relationships with healthcare professionals and offering information about emerging clinical trials. It's an education role, not sales. You can work in government affairs, where you monitor the new rules and regulations that are coming out. You can work in clinical trials and help ensure the launching studies work well. You can work in the statistical analysis of clinical trials. You can work internally to help plan publications. You can work in one of their laboratories conducting preclinical research.

Many people working in the pharmaceutical industry are purpose-driven, and motivated by the idea of developing and delivering solutions to complex diseases around the world. As a result, many people are drawn to pharma jobs due to the potential to make a positive impact on the health of people globally. Individuals often progress into senior roles in order to have an even greater impact.

Working in the pharmaceutical industry can be incredibly fulfilling. Many pharmaceutical companies are constantly pushing the boundaries of medical science. While there are many research organisations that are leading cutting-edge science, the pharmaceutical industry is an exciting place to work as well. This aspect of the industry can be particularly attractive to individuals looking for a challenging and dynamic career.

I found there were some great aspects of working for a pharmaceutical company. The atmosphere was friendly. We made sure to take a lunch break altogether. I still remember the crème brûlée for dessert. It was delicious.

Some of the skills I developed in research truly helped me make the big jump from academia to corporate life. I had done my PhD in an unfunded lab in an isolated town, so I had to learn from the Internet and teach myself. It was the same when transitioning from academia to the corporate world. A lot of the time, my boss would ask me to do something, and I wasn't sure how to do it. But combining googling, searching the internal intranet for similar projects, and asking thoughtful questions meant that I could complete my "deliverables" before the deadline.

You need to know that the pharmaceutical industry does have a bad reputation in some circles. This is because it is seen as making a profit from ill health, a lack of transparency, and questionable marketing of treatments. While this may not be everyone's experience, all the people I've worked with in the pharmaceutical industry have genuinely had the best interests of patients at heart.

The professional distance between myself and my colleagues was very different from academia. In academia, every Friday night, we would go out for beers afterwards. It was very different in the corporate world, where people did their work, and then went home to their families.

The company was also much more extensive than my lab. There were thousands and thousands of people working at this company across multiple continents. The CEO had no idea who I was. I was a teeny tiny cog in a vast machine.

I spent my years there producing epidemiological reports about rare diseases. Unfortunately, very little of my research was published. A lot of it remained buried in the depths of the files of the company intranet because it included confidential reports about pre-clinical investigations, proprietary information, and data that had not yet been approved by regulatory bodies.

After working at a pharmaceutical company for several years, I decided to leave. Why? For me, it all came down to feeling like a small cog in a big machine. I realise that I value autonomy, freedom, and agility. I wanted to work where we could move fast and quickly, and where I knew the names of every person in my organisation.

ADVANTAGES OF WORKING AT A PHARMA COMPANY

- Good work/life balance
- Purposeful work
- Generally, a positive workplace culture
- Career progression is visible and encouraged
- Good salary
- Some roles allow for travel or are field-based

DISADVANTAGES OF WORKING AT A PHARMA COMPANY

- Projects can go slowly due many levels of approvals
- Effort is needed to secure internal buy-in
- Expectations of delivering high quality work on time
- Defined projects due to strong industry regulations
- Your social media is subject to rules to ensure you are not breaching regulations
- Lack of agility
- Many management levels between you and the CEO

FILL THE GAPS IN YOUR CV ON YOUR OWN

- Get a part-time job working in a pharmacy
- Become a student member of the International Society for Pharmaceutical Engineering, the Drug Information Association, the UK Pharmaceutical Information and Pharmacovigilance Association, the Pharmaceutical Society of Australia, or the Canadian Society for Pharmaceutical Sciences.

FILL THE GAPS IN YOUR CV IN YOUR WORK

- Volunteer to work on a strategy committee
- Enter competitions for posters and presentations
- Take a course in presentation skills
- Take a class in Excel skills
- Ask to organise a meeting with external people
- Travel to congresses and give oral presentations
- Ask to manage volunteers and students
- Look for internships at pharmaceutical companies
- Subscribe to The Pharmaceutical Journal and Journal of Pharmaceutical Sciences

GREEN FLAGS IN JOB DESCRIPTIONS

- Clearly outlines the role and responsibilities
- Training and opportunities for advancement
- Competitive compensation and benefits
- The company's mission and values
- Cutting-edge research and development
- Teamwork and communication
- Reasonable and flexible working hours
- Opportunities for remote work
- "strong moral compass"
- "Strong focus on ethics and compliance"
- "Collaboration with cross-functional teams"

RED FLAGS IN JOB DESCRIPTIONS

- Emphasis on sales targets or revenue goals
- Unrealistic qualifications or expectations
- The job description is vague
- Long hours or frequent overtime

- The company has a history of ethical or legal issues
- The company has poor reviews on Glassdoor
- The company careers section is 99% sales jobs
- "Ability to work under pressure"
- "Fast-paced work environment"
- "Availability required for weekend work"
- "Experience meeting deadlines for multiple projects simultaneously"
- "Ability to work independently with minimal supervision"

THE BEST COURSES TO TAKE

Good pharmacovigilance practices

`pharmalessons.com/free-courses/gvptraining/`

Upon completion of this course, trainees will be able to understand the current legal regulations; outline the main requirements; get familiar with the essential documents and understand important trial-related files such as the pharmacovigilance system master file, the periodic safety update report and the study protocol.

Clinical trials operations specialisation

`coursera.org/specializations/clinical-trials-operations`

Learners will develop insights and build the skills they need to design, manage, and monitor clinical trials, as well as analyse, document, and communicate the results. Learners will also learn best practices regarding ethics, safety, participant recruitment, regulatory compliance, and reporting standards. The core principles and skills of the specialisation will lay the foundation for a successful career in the field.

~

SAMPLE COVER LETTER

Dear Ms Li,

I would like you to consider me for the position of Outcomes Manager as advertised on the ABC.com website. I am a scientist with years of experience.

I believe in ABC's mission to improve the lives of people with severe diseases. I feel that I am well equipped to support the newly developed Patient Solutions Teams to achieve their missions and provide targeted data to the Benefit-Risk Teams and the Benefit-Risk Board.

I have collaborated successfully with multidisciplinary teams of colleagues from Europe, Asia, and the Americas. I have presented oral and poster presentations of my scientific findings at international meetings. I have contributed to the development of Standard Operating Procedures and Scientific Advice Packages.

My PhD training in medical research has given me a strong background in rigorous scientific methodology. My Masters in Public Health allows me to perform signal evaluation analysis, risk-benefit assessments, and calculate incidence and prevalence.

With a strong background in epidemiology, and experience analysing observational studies, I am confident I am the best-qualified candidate for the position of Outcomes Manager at ABC Pharma.

Thank you for your consideration.

Sam Jones.

~

SAMPLE CV

Sam TANG, MPH, PhD

Scientist with a background in medical research. Skilled in the management and motivation of diverse research teams and specialising in the investigation of clinical trials and observational data for health outcomes and safety signal detection.

EXPERIENCE

Senior post-doctoral fellow

Unseen University Medical Center, Springfield, Eddis

- Collaborated with clinicians
- Presented findings at 3 international symposia
- Conducted 15 meta-analyses on clinical trial data
- Drafted 6 compliance protocols
- Authored a widely cited paper on T-cell response factors
- Managed a team of 6 PhD students

EDUCATION

Expert Workshop in Economic Evaluation in Health Care

University of Arendelle, Arendelle

- Focused on costing methods and utility measurement
- Collaborated on a group project analysing healthcare costs for 2 case studies.
- Completed 10 hours of interactive sessions on decision analytic modelling techniques.

L. E. MAKAROFF

Doctor of Philosophy in Immunology

Banting University, Gondor

- Authored 2 papers published in peer-reviewed journals
- Presented findings at 6 national and international seminars, receiving 2 best presentation awards.
- Conducted 3 seminars on advanced immunological techniques for postgraduate students.
- Organised and moderated a symposium , attended by over 100 delegates.

Graduate Award in Statistical Methods for Research in the Life Sciences

Banting University, Gondor

- Conducted a detailed multivariate analysis project, interpreting data from 3 complex clinical studies.
- Completed 40 hours of intensive training in advanced statistical methods.
- Received a distinction for a thesis focusing on applications of statistical methods in research.

SELECTED PUBLICATIONS

Wang FF, Huang G, Zhang KA, Chen S, Tang SE. Compliance with pharmacotherapy and healthcare costs: a retrospective claims database analysis. *Applied Health Economics and Health Policy.* 2112;2(2):62-72

Tang SE, Li A, Liu C, Wang F. Disorders in Disease: Prevalence and Health Outcomes in a US Claims Database. *Journal of Disease.* 2111;1(1):65-74

SKILLS

Environments

Windows, Mac OS X

Statistics

SAS (Base / Stat / Graph), S-plus, R, SPSS, Stata, JMP

Applications

MS Office, Statistica, MS Access, Datavision, CEWorks, SÆftyWorks, Microsoft Teams

~

SUMMARY

- The pharmaceutical industry offers various types of jobs, including advocacy, medical science liaison, government affairs, clinical trials, statistical analysis, and writing roles.
- Advantages of working in the industry include work-life balance, knowing you're helping people, supportive workplace culture, career progression, good salary, and potential for travel or field-based roles.
- Challenges include adjusting to a different corporate culture, navigating regulatory constraints, and feeling like a small part of a large organisation.

~

CHAPTER 11
BIOTECH START-UP

I am not afraid of storms, for I am learning how to sail my ship.

LOUISA MAY ALCOTT

Joining a small biotech start-up will be a dive into a vibrant and challenging world. Unlike the structured environment of prominent corporations, a start-up is a dynamic space where adaptability and initiative are essential. You will be thrown into a whirlpool of new terms and concepts.

In this environment, job roles are often fluid and ever-evolving. You could be involved in a range of activities These activities could range from lab work, to engaging with patient groups to raise disease awareness, to liaising with healthcare professionals about clinical trials.

In small biotech, there's a sense of purpose in addressing unmet medical needs with innovative solutions. This desire to make a tangible difference attracts many to the field, often leading them to take on multiple roles or advance quickly to take on more responsibility.

The pace is fast, and the stakes are often high. You will be part of a small team, tackling significant challenges and pushing the boundaries of scientific understanding and application. This dynamic environment could be perfect for those seeking a challenging, hands-on career.

The culture in a small biotech start-up is often noticeably different from that of a large pharmaceutical company. The atmosphere is often more informal and collegial. When you meet face-to-face, lunch breaks are likely to be a time for team bonding, often leading to brainstorming sessions. The social dynamics will be more intimate than the distant professional relationships in big corporations.

The scale of operations in a start-up is vastly different. In contrast to a large company's anonymity, everyone knows your name. Your contributions will be visible and valued, and there's a real sense of being part of something significant, even if it's small in size.

Your research skills, honed during your studies, will prove invaluable. In a start-up, resourcefulness is vital. You will learn on the go, using the Internet or consulting colleagues to navigate unfamiliar tasks and meet tight deadlines.

Biotech start-ups don't always have the best reputation and are often viewed as risky or unstable. The nature of projects in such an environment can be volatile, often undergoing sudden shifts in direction due to the ebb and flow of funding or unexpected research outcomes. Managing internal stakeholders, while typically less formalised than in larger corporations, remains a crucial aspect of the job and requires a deft touch. Employees are frequently under pressure to deliver high-quality work within stringent timelines, a demand that can be as exhilarating as it is demanding.

In smaller companies, the rules and regulations can differ from those in bigger firms, and they can be tricky to handle. It's important to understand and follow these rules to ensure your company stays compliant while moving forward. These regulations can change often, and you will need to stay up to date.

In a small start-up, you often have fewer resources than big companies. This means you need to be really smart and creative in using what you've got to get the best results. This also means every small success can have a big impact and bring a real sense of achievement.

Joining a small biotech start-up offers a dynamic and challenging experience, contrasting the structured environment of larger corporations. These start-ups are characterised by fluid job roles, rapid pace, and high stakes, where contributions are more visible and impactful. The atmosphere is informal and collegial, fostering close-knit team dynamics. Employees often wear multiple hats, adapting to varying tasks from lab work to liaising on clinical trials. Start-ups typically focus on innovative solutions to unmet medical needs, although they can be seen as risky due to funding volatility and resource limitations. Despite these challenges, working in a start-up can be exhilarating for those seeking a hands-on, impactful career.

ADVANTAGES OF WORKING AT A BIOTECH START-UP

- A sense of purpose and direct impact on projects
- A collaborative and close-knit team environment
- Opportunities for rapid growth and learning
- Good work-life balance despite the fast-paced nature
- Often opportunities to work remotely

DISADVANTAGES OF WORKING AT A BIOTECH START-UP

- Projects can be subject to sudden changes
- It can be challenging to distance yourself from harsh personalities due to small team sizes
- Unstructured job roles
- Expectations to deliver work under tight timelines
- Regulatory hurdles
- Limited resources

FILL THE GAPS IN YOUR CV ON YOUR OWN

- Working part-time in a related field
- Pursue internships in biotech firms
- Joining relevant professional societies
- Attending industry conferences and networking events
- Start a science blog or podcast
- Engage with online communities to discuss current challenges in biotech
- Learn biotech-related software, coding, or lab techniques through self-study

FILL THE GAPS IN YOUR CV IN YOUR WORK

- Participate in strategy-focused committees
- Engage in competitions and presentations
- Develop strong presentation and analytical skills
- Seek roles involving external collaboration
- Stay informed through industry-specific publications

GREEN FLAGS IN JOB DESCRIPTIONS

- Clear role definitions and advancement opportunities
- Ethical work practices are emphasised.
- Collaboration opportunities are outlined
- Reasonable work hours and flexibility
- Share options scheme
- The company has good reviews on Glassdoor

RED FLAGS IN JOB DESCRIPTIONS

- Overemphasis on aggressive goals
- Vague job descriptions and unrealistic expectations
- Company has a history of legal issues
- Excessive jargon or buzzwords that obscure role clarity
- Advertised through a recruiting agency without stating the name of the start-up
- "Rockstar" or "Ninja" used to describe the role
- "We'll give you responsibility from day one"
- "Thrives balancing many tasks at once"

THE BEST COURSES TO TAKE

Google data analytics: Professional certificate

```
coursera.org/professional-
certificates/google-data-analytics
```

Designed to be completed in under six months with no prior degree or experience required. The program focuses on teaching skills such as data cleaning, analysis, and visualisation using tools like spreadsheets, SQL, R programming, and Tableau. Students will learn through practical, real-world scenarios to develop skills in creating effective project documentation and analysing and visualising data.

Google project management: Professional certificate

`coursera.org/professional-certificates/google-project-management`

This program, developed by Google, teaches that prepare learners for entry-level roles in less than six months, without requiring a prior degree or experience. The course covers an understanding of project management practices and skills, including creating project documentation, understanding Agile project management, and developing skills in strategic communication and stakeholder management.

~

SAMPLE COVER LETTER

Dear Dr Carter,

I am writing to express my interest in the Junior Research Scientist role at TechTech Innovations, as advertised on BioTech Careers. As a graduate with a degree in Biochemistry from Starling University, I am eager to apply my passion for biotechnology to contribute to your innovative team.

While at Starling University, I engaged in several projects that aligned closely with the work at TechTech Innovations. For instance, my senior thesis focused on the novel use of CRISPR technology in gene therapy, which required extensive research, data analysis, and collaboration. This experience honed my technical skills in bioinformatics and cultivated my ability to work effectively in team-driven environments.

I am drawn to TechTech Innovations because of its commitment to pioneering new genetic therapies for complex diseases. I am especially interested in your ongoing project on targeted gene therapy for cystic fibrosis, and I am confident that my background in genetic research will be a valuable asset in this area.

I have developed strong communication, problem-solving, and leadership skills through my internship at BioSolutions Ltd. My role as a Research Intern involved coordinating with cross-functional teams to streamline lab processes.

I have consistently demonstrated a commitment to continuous learning and professional development. My proactive approach includes staying updated with the latest advancements in biotechnology through attending webinars, participating in workshops, and being an active member of the Young Biotech Researchers Forum.

I possess a keen interest in interdisciplinary collaboration, understanding that breakthroughs in biotechnology often occur at the intersection of multiple fields. During my time at Starling University, I actively participated in joint projects with the departments of Computer Science and Engineering, gaining valuable insights into how technology can augment biological research.

My ability to quickly adapt to new challenges and integrate innovative solutions makes me well-suited to the dynamic and forward-thinking environment at TechTech Innovations. I am eager to bring my passion for biotechnology and my drive for innovation to your company.

Thank you for considering my application.

I look forward to the opportunity to discuss how my skills and enthusiasm align with the needs of your team.

Sincerely,

Claire Deller

SAMPLE CV

Claire Deller

Eager to apply my expertise in biotechnology to a dynamic start-up environment. Passionate about innovative solutions to complex biotechnological challenges.

EXPERIENCE

Research Assistant

Kingswell University Biotech Lab, Kingswell University

- Conducted research on targeted gene therapy techniques, achieving a 15% increase in the efficiency of non-invasive delivery methods
- Demonstrated proficiency in Leibig condensers and Van De Graaf generators
- Collaborated with a 5-member research team to design and analyse complex Miller-Urey experiments

Biotech Intern

BioFuture Innovations, Middlemarch

- Played a pivotal role in developing a novel diagnostic tool for early detection of genetic disorders, contributing to a 30% reduction in detection time
- Participated in phases of the project lifecycle, from initial concept to data analysis, across 3 project milestones
- Acquired an understanding of industry-specific practices, including regulatory standards

Volunteer Coordinator

Peterswood Community Health Initiative

- Coordinated and executed 12 health awareness campaigns and 8 workshops annually, resulting in a 40% increase in community engagement
- Mobilised and managed a team of 30 volunteers, enhancing the efficiency and reach of each event
- Implemented engagement strategies that led to a 25% increase in repeat volunteer participation

EDUCATION

Master of Science in Biology

Kingswell University, Peterswood

- Specialised in molecular genetics, achieving a distinction for thesis on gene expression patterns.
- Completed coursework with a focus on advanced biostatistics and bioinformatics analysis.
- Actively participated in cross-departmental seminars, enhancing interdisciplinary collaboration skills.

Certifications

- Google Project Management
- Bioinformatics for Genomics
- Drug Development Product Management

Skills

Technical Skills: Jacob's laddering, Tesla coiling, Lissajous patterning

Software : MS Office, R, Python, Tableau

Soft Skills: Teamwork, Communication, Time Management

Hobbies

Amateur radio operating, bird watching, 3D printing

$\sim$

SUMMARY

- **Working in a small biotech start-up is a world where your work can directly contribute to ground-breaking scientific advances.**
- **The pace is fast, the environment is dynamic, and the sense of community is tangible.**
- **While it may lack the stability and resources of large pharmaceutical companies, a start-up offers an opportunity to be at the forefront of innovation in the biotech industry.**

CHAPTER 12
CONSULTANCY COMPANY

*Melbourne is 11 hours ahead of London, but that doesn't
make London slow. Everyone is running their race in
their own time. They are in their time zone, and you are
in yours. You're not late. You are very much on time.*

ADAPTED FROM "ON TIME" BY
FAYE CHEN

Consultancy companies can be a great option immediately after your PhD or if you've been stuck in a lab for a while and feel like you have no transferable skills. They are always looking for fresh blood. You are much more likely to get a consultancy job than any other option.

Consultancy companies will use you up and spit you out. That said, they are likely to hire you if you're looking for a job. In my experience, there is very much a churn mentality when it comes to consultancies. Many consultancies are known to have very toxic cultures. There is a significant turnover, with minimal support. They need a lot of staff to deliver many projects. They will overpromise their clients and try to deliver multiple projects with insufficient staff.

If you join a consultancy, suddenly you will find yourself given vast amounts of responsibility, crazy tight deadlines, and very little support. It will be a stressful time. You'll have to work long hours to deliver. You'll have to deal with clients who are frustrated that your work isn't good enough. You will have to work twice the amount of time billed to complete the work. You will have to research everything on your own and do everything on your own.

It is easy to be caught up in the stress of it all. To define yourself as working hundred-hour weeks and working through the night to meet an insane deadline.

But once again, despite the downsides, this is a great stepping stone. This is the place to learn. This is the place where you will get direct contact with industry partners, charity partners, and academic partners. You're going to get hands-on experience and contact with so many different industries and projects. You're going to have a much better idea of what you want to do with your life and what the world is like out there.

I suggest only planning to work for a consultancy for up to two years. Twelve months into your consultancy, you should have tried many different projects and interacted with many clients. Now that you have a firm idea of what you want to do and your CV looks much better than it did a year ago, you can start planning your next steps.

You can also ask your consultancy to pay for all training courses. For example, if a PRINCE2 project management course is on offer, take that.

Your job at the consultancy is to leave it with a fantastic CV and robust skill set, so you have the knowledge — and the proof of that knowledge — to find a career path that is a better match for you.

Don't try to make partner.

Don't try to climb the ladder.

I've never met anyone truly happy in a consultancy unless they are the founder of the consultancy.

Maybe I'm wrong. If you love working as a Consultant and have worked there for over 20 years, then please send me an email and let me know, but sadly all the consultancies I have seen have not been good for anyone's mental health. The best and brightest tend to leave swiftly and move onwards.

But as I said earlier, consultancies can be great for networking, learning, and intense real-world experience.

ADVANTAGES OF WORKING IN CONSULTING

- Exposure to a variety of industries
- Rapid skill development
- Easy to find a job
- You will learn a lot
- You will be given a lot of responsibility
- You will make many connections

DISADVANTAGES OF WORKING IN CONSULTING

- Long hours
- Little support
- Competitive work environment
- Low salary for entry-level roles
- Pressure to meet unrealistic deadlines

FILL THE GAPS IN YOUR CV ON YOUR OWN

- Take a course in presentation skills
- Take a class in Excel skills
- Start an X, Mastodon, or LinkedIn account

FILL THE GAPS IN YOUR CV IN YOUR WORK

- Work on a strategy or budget committee
- Write a short monthly report of your progress
- Enter competitions for posters and presentations
- Draft the press release related to your publications
- Contribute to the newsletters of your organisation
- Ask to organise a meeting that involves external people
- Travel to external congresses and give oral presentations
- Ask to manage volunteers and students

GREEN FLAGS IN THE JOB ADVERTISEMENT

- Salary is stated
- Additional pension contributions are stated
- Generous vacation time
- Flexible working is encouraged
- Expected weekly working hours are stated

RED FLAGS IN THE JOB ADVERTISEMENT

- "We work hard and play hard."
- "We are a family."
- "ability to work under pressure"
- "entrepreneurial culture"
- "push the limits of what's possible"
- "work is challenging and exciting"

THE BEST COURSES TO TAKE

Google project management: Professional certificate

`coursera.org/professional-`
`certificates/google-project-management`

This program, developed by Google, teaches that prepare learners for entry-level roles in less than six months, without requiring a prior degree or experience. The course covers an understanding of project management practices and skills, including creating project documentation, understanding Agile project management, and developing skills in strategic communication and stakeholder management.

Introduction to Management Consulting

`coursera.org/learn/introduction-to-`
`management-consulting`

During this 7 hour course, participants will learn what management consultants do, how to learn consulting skills, and earn a shareable career certificate.

PrepLounge case interview basics

`preplounge.com/en/case-interview-basics`

Knowledge and methodologies that enable you to prepare for a consultancy interview. It includes an overview of the interview in general, concrete advice about specific case types, business concepts, and common business terms.

SAMPLE COVER LETTER

Dear Ms Arnott,

I was thrilled to discover your website advertisement for a junior consultant position, and I am excited to apply for the role. As an experienced immunologist with a passion for scientific communication, I am confident that my skills and expertise make me a perfect candidate for the job.

With over six years of experience in delivering award-winning presentations, writing journal articles, and constructing protocols, I have developed a strong reputation as a detail-oriented scientist. As a native English speaker, I possess excellent communication skills, and I have a PhD in Immunology, a Bachelor of Science (Honours) and a Graduate Award in Scientific Communication. I am also pursuing a Master's in Public Health.

Throughout my career, I have designed and managed multiple scientific projects while supervising junior research scholars. I am proficient in statistical analysis and packages such as R, SAS, SPSS, and S-Plus, and I am dedicated to delivering exceptional reports to clients under a strict deadline. You can find further details of my experience in the attached CV.

I am excited about the opportunity to relocate to Genovia and work for Business Consultancy Partners.

Thank you for considering my application. I look forward to hearing back from you soon to discuss my suitability for the role.

Sincerely,

Dr Tang

~

SAMPLE CV

Sam TANG

Exceptional expertise in immunology that has significantly advanced patient care. Master's in International Public Health and a PhD in Immunology complemented by a strong skill set in project management, clinical epidemiology, advanced statistical analyses, and comprehensive knowledge in therapeutic areas.

EXPERIENCE

Unseen University Medical Center, Springfield, Eddis

Senior post-doctoral fellow

Designed strategies reducing patient autoimmune complications by 30%, authored 5 scientific articles in top-tier journals, conducted 3 meta-analyses and and drafted 6 protocols.

EDUCATION

Masters of International Public Health

Worthington University, Gondor

Clinical epidemiology, medical event databases, patient assessment, treatment identification, phase I to IV clinical trials, systematic reviews, and meta-analyses.

Doctor of Philosophy in Immunology

Banting University, Gondor

Demonstrated a novel role for chromosome proteins to produce T cells. Improved immunological protocols, utilised advanced statistical analyses, and managed data sets.

THERAPEUTIC AREAS

- <u>Diabetes</u> – investigated the role of CD4 and CD8 T cells in the suppression and activation of Type I diabetes. Determined a novel role for T cells in autoimmunity.
- <u>Infectious Diseases and Vaccines</u> – designed vaccination schedules for viral and bacterial diseases. Discovered a role for T cells for optimal memory response.
- <u>Oncology</u> – conducted research into the stages of T cell maturation that can lead to lymphoma. Characterised a protein that can cause a decrease in lymphocyte number.

<u>Epidemiology and statistics</u>

- Medical event databases
- Audits, quality, and safety
- Healthcare information management
- Survey design and analysis
- Multivariate analysis and classification
- Principal components analysis
- Linear regression
- Generalised linear models

IT Skills

- Environments: Windows, DOS, Linux, MacOS
- Languages: SAS (Stat / Graph), S-plus, R, SPSS, Stata
- Databases: MS Access, SQL
- Tools & Applications: MS Office, Statistica

Hobbies

Beekeeping, blacksmithing, podcasting

~

SUMMARY

- **Consultancies offer a viable career option after a PhD, providing job opportunities and exposure to various industries, but they often have toxic cultures and high turnover rates.**
- **Working in a consultancy entails significant responsibility, tight deadlines, and minimal support, leading to stress and long working hours.**
- **Use your time in a consultancy to gain valuable experience, network with industry professionals, and build a strong skill set, with the intention of transitioning to a more suitable career path within two years.**

CHAPTER 13
MEDICAL DOCTOR

As long as you're learning, you're not failing.

BOB ROSS

The chief benefit of getting medical training after a PhD is that you'll get to see what clearly does and doesn't work on both sides, and you have the role of bridging the gap. The best-suited person for a medical doctor role is someone who has strong communication skills, the ability to work well under pressure, empathy, a passion for helping others, and a commitment to lifelong learning.

There will be a great deal of learning. The sheer amount of knowledge you need to take in and memorise, often in a short period, can be staggering. And the stakes are high since you're not just memorising this information for the sake of your exams, but also for the people you will be treating. A good memory and an ability to memorise facts quickly are convenient. Only some medical students have a great memory, of course, in which case long hours and repetitive reviewing come into play.

As someone studying to become a medical doctor, your path will be 90% defined for the next eight to ten years. No matter how well you do in school, how creative you are, or how many medical mysteries you solve, you will need to start at the bottom of the hospital hierarchy and move up. This holds true in medical school, residency and fellowship.

Medicine is a technical discipline. People get sick in many ways, and illnesses manifest slightly differently for everyone. The only thing constant is the variability; you need to be able to adapt your knowledge to this variability.

You treat patients, not conditions, but dealing with human beings is complicated. This means it's ideal to be empathetic, but you also need to look out for yourself, be strong, and compartmentalise. You witness people going through a lot of pain. You see people losing their dignity and sometimes cannot alleviate their distress as this happens. This can also be distressing to you, the physician, and learning how to deal with it takes some time.

In this field, you must understand that perfectionism is unattainable. Most medical doctors have a type A personality and find it hard to accept that good enough is good enough. Repetitive tasks like outpatient clinics can be tedious. And it's a lot of hard work talking to patients. However, it provides excellent patient service, and you learn a lot. Remember that the enemy of good is perfect.

If you decide to enter this field, it's important to set boundaries. You don't have to always be all things to all people. Prioritise your goals professionally and personally. Learning to say no has been a potent and effective tool; do it by looking at the request and seeing whether it aligns with your goals.

Mentorship is vital in medicine and can come from all directions. Choose someone whose ideals and standards align with your own. It doesn't have to be a senior colleague; it can be your peers. It can be the healthcare workers around you. Nurses have been there, seen it before, and understand the politics. It's ideal to have different mentors for different goals, so sometimes it's for the clinical outcomes, sometimes it's for research, and sometimes it's just for navigating the department's politics.

Being flexible and being able to change your focus is vital. The tremendous benefit of clinical education is that it provides you with variety. It's an opportunity to shape the future of medicine. There will be seasons with ebbs and flows, so at some point. Sometimes you might focus on education; other times, you might conduct more research.

The best parts of being a doctor include the opportunity to make a significant positive impact on people's lives, the ability to work with a diverse range of patients and colleagues, and the intellectual challenge of diagnosing and treating complex medical conditions. You will have a stable salary that can support a reasonable lifestyle.

Factors to consider when deciding whether to pursue a medical doctor career include the time and financial commitment required for medical school, the intense workload and responsibility that comes with the job, and the potential for work-life balance challenges.

ADVANTAGES OF WORKING IN MEDICINE

- You can help to alleviate pain and suffering
- It's a respected profession
- There's a wide choice of specialities
- Opportunities to teach, research, and manage
- Practical and intellectual challenges
- Reasonable salary, and this increases with experience
- Doctors will always be in demand

DISADVANTAGES OF WORKING IN MEDICINE

- The many years of study and training
- The long and challenging training
- Your poor work-life balance
- The risk of litigation
- The demanding and stressful work
- Requires continuous education and learning
- Exposure to infectious diseases

PREPARE FOR MEDICINE ON YOUR OWN

- Volunteer with a local health clinic or hospice
- Engage in community health initiatives
- Take courses on interpersonal communication
- Get a part-time job as a medical scribe, pharmacy technician, or hospital receptionist
- Join the American Medical Association, British Medical Association, Australian Medical Association, Canadian Association of Medical Education, or the Canadian Society for Clinical Investigation.

PREPARE FOR MEDICINE AS PART OF YOUR WORK

- Volunteer to work on a medical committee
- Ask to get first-aid qualified
- Choose projects that involve hospitals
- Choose projects related to medicine and healthcare
- Take courses in anatomy and pharmacology

THE BEST COURSE TO TAKE

So You Want To Be A Surgeon?

`edx.org/course/surgery`

What do surgeons do? What conditions require surgical care? In this 6-week course, you will be guided through 15 surgical specialities by a team of 40+ clinical experts. You will have a chairside view of the day-to-day work life of a surgeon and learn about diagnosis, treatment planning, and surgical care and see surgeons work in a multidisciplinary team to offer the best treatment for the patients.

~

SUMMARY

- **Pursuing a medical doctor career after a PhD offers the advantage of bridging the gap between academia and clinical practice and seeing what does and doesn't work on both sides.**
- **The advantages of being a doctor include making a positive impact, working with diverse patients and colleagues, intellectual challenges, reasonable salary, and high demand for doctors.**
- **Disadvantages include the long and challenging training, poor work-life balance, risk of litigation, demanding and stressful work, and the need for lifelong learning.**

CHAPTER 14
STATISTICIAN OR CLINICAL DATA MANAGER

*Education is the passport to the future, for tomorrow
belongs to those who prepare for it today.*

MALCOLM X

What was your favourite part of your PhD or post-doc? Was it sitting down in front of your computer and crunching those numbers? Did you spend a fair amount of your day lost in SPSS, R or detailed Excel macros? A job as a statistician or clinical data manager may be for you.

A person who is detail-oriented, enjoys problem-solving, and has strong analytical skills would be well-suited to a career as a statistician or clinical data manager. Strong communication skills are also important for effectively communicating complex data to others.

Your background in understanding where the data comes from will set you apart from pure statisticians. Statisticians may work in academia or industry, while clinical data managers typically work in healthcare or pharmaceutical companies. You can work in academia, the pharmaceutical industry, or the third sector. See those chapters for more information about the advantages and disadvantages of each industry.

A Clinical Data Manager works to plan, develop, and implement clinical data management systems. You might work closely with clinical teams to collect data accurately and efficiently. You will likely also ensure that information is appropriately stored and accessible for analysis. If you are interested in healthcare and medical research, clinical data management may be a good fit for you.

A Statistician works to help in the creation of data and the proper use of mathematical methods to develop and analyse it. You may use your skills to design experiments and surveys and then compile and interpret the collected data. You may also create new statistical analysis methods or help improve existing processes. If you enjoy working with data in a broader context, statistics may be a better fit.

The best part of working as a Statistician or a Clinical Data Manager is that you will get to work with different types of data and help find trends in them. You will be working with interesting data and solving complex problems, as well as having the opportunity to make a significant impact on research and patient outcomes. You will get to work with various people and help them understand the data.

Challenges of working as a Statistician or a Clinical Data Manager include ensuring the accuracy and completeness of data, managing timelines and expectations, maintaining data quality, overseeing compliance with regulatory guidelines, coordinating with multiple stakeholders, training staff, and dealing with complex data sets. You will potentially not be able to speak about the details of your findings due to ethical issues related to data privacy and confidentiality.

ADVANTAGES OF BEING A STATISTICIAN

- No lab work for you, but you can stay in research
- You will be in high demand
- The salaries are good
- You will develop transferable skills, providing opportunities for career growth and flexibility
- You can help shape healthcare and research

DISADVANTAGES OF BEING A STATISTICIAN

- Can be isolating
- Not many opportunities for creative expression
- A high level of technical skill and attention to detail
- You will be dependent on other people for data
- Most people will not understand the work you do

FILL THE GAPS IN YOUR CV ON YOUR OWN

- Join your local R users group
- Subscribe to the Journal of Clinical Data Management and Journal of the American Statistical Association
- Join an organisation such as the American Statistical Association, Royal Statistical Society, Statistical Society of Australia, or Statistical Society of Canada
- Get a part-time job as a data-entry clerk, maths tutor, or freelance data analyst

FILL THE GAPS IN YOUR CV IN YOUR WORK

- Take programming courses in R, SAS, and Python
- Move away from Excel
- Use LaTeX for your documents, including your CV

GREEN FLAGS IN JOB DESCRIPTIONS

- Flexible working hours and the option for remote work
- A culture of learning and growth mentioned
- Strict ethical standards in data management
- Data privacy and integrity mentioned
- "Ongoing training and development programs"

RED FLAGS IN JOB DESCRIPTIONS

- Undefined role boundaries
- "Must be able to handle several projects simultaneously"
- "Must be available to work outside of office hours"
- "Responsibilities may extend beyond typical tasks"
- "Looking for a self-starter who can operate effectively"
- "Ever-changing work landscape"
- "Flexible approach to data handling needed"

THE BEST COURSES TO TAKE

Data Science: Foundations using R Specialisation

`coursera.org/specializations/data-science-foundations-r`

This Johns Hopkins University and SwiftKey specialisation covers foundational data science tools and techniques, including getting, cleaning, and exploring data, programming in R, and conducting reproducible research. Participants will learn to use R to analyse data, how to set up R-Studio, how to use a GitHub repository to manage data science projects, write up a reproducible data analysis using Knitr, and publish reproducible web documents using Markdown. There is also the opportunity to earn a career certificate.

L. E. MAKAROFF

Questionnaire Design for Social Surveys

`coursera.org/learn/questionnaire-design`

This course will cover the basic elements of designing and evaluating questionnaires. It will review the process of responding to questions, challenges and options for asking questions about behavioural frequencies, practical techniques for evaluating questions, mode specific questionnaire characteristics, and review methods of standardised and conversational interviewing.

~

SAMPLE COVER LETTER

Dear Ms Devi,

I am responding to your advertisement for a Clinical Data Manager published on your website after a referral from Prof Yang of Springfield University.

I have spent many years at the bench working on the fundamentals of the immune system. I am now interested in using the skills I have developed to improve the speed at which therapeutics can be integrated into treating patients with diseases. I have been very impressed by the quality of clinical trials organised by ABCDE, and I would value the opportunity to be involved in this organisation.

I am currently working as a PhD research student at the Unseen University, Springfield. I have experience in dataset management and analysis in the medical and psychological fields. I have conducted principal components analysis of data ranging from microarray expression assays to population studies. I have published rigorous analyses of my experimental studies in high-impact journals and communicated these results to colleagues at international conferences. My scientific studies demonstrate that I am capable of careful observation, record-keeping, and productive multidisciplinary collaboration with other researchers.

I also have extensive experience using statistical packages, such as R, SAS, SPSS, S-Plus, and Prism. I am motivated, enthusiastic, and dedicated to achieving experimental design and analysis excellence. Please see my attached CV for further details.

I bring a proactive and detail-oriented approach to all aspects of clinical data management. My ability to efficiently manage large datasets and ensure their accuracy has been a key component in my research. Furthermore, I possess strong communication skills, which would allow me to effectively liaise with clinical teams and ensure seamless integration of data management strategies.

My commitment to quality and precision makes me confident in my ability to contribute significantly to the ongoing success of your clinical trials at ABCDE. I am particularly excited about the prospect of applying my skills in a dynamic and impactful setting, and I am eager to contribute to the innovative work being done at your organisation.

I can be reached via email at tang@u.unseen.edu or via phone at +1 206 240 6291.

I hope to hear from you about this employment opportunity.

Yours sincerely,

Sam Tang

SAMPLE CV

Sam Tang, Ph.D.

Masters of International Public Health. The Worthington *Sunnydale University*, Gondor. In progress –completion in 2129.

- Epidemiological study of incidence and risk
- Sensitivity and specificity of diagnostic tests
- The application of statistical process control charts

Doctor of Philosophy in Biochemistry and Molecular Biology. Gondor National

Unseen University, Gondor, 2124-2127.

- Demonstrated a novel role for chromosome proteins in T cell development
- Produced manuscripts that were accepted for publication in top journals
- Taught classes in medical ethics, microbiology, pathogens, and immunology

Graduate Award in Statistical Methods for Research in the Life Sciences.

Unseen University, 2125.

- Earned the top score in survey design
- Conducted multivariate analysis, linear regression, and generalised linear models.

PUBLICATIONS

Sam E. Tang and Mary J. Silva. 2129. Maturation influences the T cell response. *Proceedings of the National Academy of Sciences Eddis,* in press.

Sam E. Tang and Rita Devi. 2128. Genetic control of T cell tolerance. *Genetic Predisposition to Disease,* pp 39-71. Cliffport: Avan Publishers.

SKILLS

R, SAS, SPSS, S-Plus, Prism

∿

SUMMARY

- A career as a statistician or clinical data manager is well-suited for individuals who enjoy working with numbers, have strong analytical skills, and are detail-oriented.
- The advantages of working in these roles include working with interesting data, making a significant impact on research and patient outcomes, and developing transferable skills.
- Challenges include ensuring data accuracy and compliance, managing timelines and expectations, and dealing with complex data sets.

CHAPTER 15
TEACHER

One child, one teacher, one book, and one pen can change the world.

MALALA YOUSAFZAI

Embarking on a career in teaching offers an enriching and impactful journey. As a teacher, you contribute to the intellectual and personal development of students, engaging with a diverse community of learners and fellow educators. The tangible impact of your work is seen in the growth and achievements of your students.

A day in the life of a teacher is filled with opportunities to shape young minds and hearts. From explaining complex theories to exploring the depths of literature, teachers bring academic subjects to life, making them accessible and exciting. The joy of seeing a student grasp a difficult concept, the pride in their accomplishments, and the ability to influence their future paths are among the most gratifying aspects of teaching.

Working in education often means being part of a larger team in a school or educational institution. This environment can present administrative challenges but also provides access to a wide array of resources and professional development opportunities. Teachers are driven by their school's mission to educate and empower.

Professional development in a teaching career is multifaceted and essential for your growth and the enhancement of student learning. It typically involves participating in workshops which provide you with new strategies, tools, and knowledge to bring back to your classrooms. Pursuing further qualifications, such as advanced degrees or specialised certifications, is another key aspect, allowing you to deepen your expertise in subject areas or educational methodologies. Collaboration on educational initiatives, such as curriculum development or school reform projects, also forms a critical part of professional development. These collaborations can be within your own school, with other institutions, or through regional or national educational organisations.

Teaching requires proficiency in various educational technologies, curriculum design, classroom management, and assessment strategies. Skills in digital literacy, online engagement platforms, and data analysis for student performance are increasingly important. The ability to adapt to diverse learning environments and to tailor teaching methods to meet individual student needs is crucial.

Teachers have the unique opportunity to extend their impact beyond the classroom by leading or sponsoring after-school clubs. These clubs can be a reflection of their personal interests and values, allowing them to connect with students on different levels. For instance, teachers passionate about social justice might lead a human rights club, where students learn about and engage in activities that promote equality and fairness. Those with a penchant for technology might sponsor a robotics club, introducing students to the world of STEM in a hands-on, collaborative environment. For those who value imaginative play and strategic thinking, leading a Dungeons and Dragons club can offer an engaging outlet, fostering teamwork and problem-solving skills in a fantastical setting. These after-school activities not only enrich students' educational experiences but also strengthen the school community by bringing together individuals with shared interests and aspirations.

Challenges in teaching include navigating complex educational systems and policies, managing diverse classroom dynamics, and often working within limited resource settings. The nature of teaching is demanding, requiring a high level of commitment, patience, and adaptability.

Strikes have become a notable aspect in some regions, reflecting broader issues within the education sector, including pay disputes and working conditions. While school shooting drills are not a common practice in most countries around the world, the very notion underscores the growing concerns around safety in educational environments globally.

Respect for the teaching profession has also been a point of contention. Teachers often report a lack of societal respect for their role, which can be demoralising and contribute to a feeling of undervaluation. This issue is compounded by the high rates of burnout in the profession, attributed to long hours, high-stress environments, and the emotional demands of working closely with students.

Furthermore, budget constraints in many educational systems lead to teachers often having to use their own money to purchase school supplies. This financial burden is a significant strain, reflecting the wider issue of underfunding in the education sector. These challenges highlight the need for systemic changes to better support teachers and enhance the teaching and learning environment.

ADVANTAGES OF BEING A TEACHER

- Annual leave is the school holidays
- Make a lasting impact on students
- Variety and dynamism of work
- Continuous learning
- Flexibility in hours, usually between 7:30 am to 4:00 pm

DISADVANTAGES OF BEING A TEACHER

- Cannot take holidays during term time
- Workload pressures
- Ongoing professional development
- Challenging admin
- Navigating reforms and policies
- Low salary

FILL THE GAPS IN YOUR CV ON YOUR OWN

- Write articles and blog posts about teaching
- Volunteer at local schools or community centres
- Engage in online educational forums
- Create a portfolio of your educational materials

FILL THE GAPS IN YOUR CV IN YOUR WORK

- Take on teaching assistant responsibilities if available
- Lead training workshops
- Mentor junior researchers
- Present your research at conferences
- Get involved in committees within your institution
- Contribute to the creation of course content
- Help to revise your institution's curriculum

- Participate in student affairs
- Volunteer to improve educational technology

GREEN FLAGS IN JOB DESCRIPTIONS

- Clear school ethos
- Clear staff roles
- Clear salary details
- Work-life balance support
- Professional development support
- Financial assistance with moving expenses

RED FLAGS IN JOB DESCRIPTIONS

- Vague responsibilities
- Required to provide instruction in multiple subject areas
- No support mentioned
- Expectation to provide unpaid overtime
- "Enforces Board policies, regulations, and rules"
- "Secures and maintains school property and materials"

THE BEST COURSES TO TAKE

Making sense of learning to teach

www.open.edu/openlearn/education-development/learning-teach-making-sense-learning-teach/

The Open University's free course is the first in a four-part series focused on the journey of becoming a teacher. It delves into various approaches to teacher education and routes into teaching, helping learners to understand the philosophical and practical differences between these approaches. The course is rooted in research about students' experiences of learning to teach. It discusses the implications of designing teacher education programmes. Key learning outcomes include understanding different perspectives on Initial Teacher Education, the role of student teachers ITE, and effective student teacher learning.

Foundations of teaching for learning: Being a teacher

`coursera.org/learn/being-a-teacher`

This course offered by the Commonwealth Education Trust on Coursera is designed for individuals who are currently teaching or aspire to teach in various settings. It includes six modules covering a range of topics essential for educators, such as teaching methodologies, reflective practice, pedagogy, and evaluation. The course takes approximately 11 hours to complete over a three-week period, allowing learners to study at their own pace. Upon completion, participants can earn a shareable certificate, which can be added to their LinkedIn profile or resume.

~

SAMPLE COVER LETTER

Dear Principal Doe,

I am writing to express my enthusiasm for the teaching position advertised at Springfield Academy. As a recent Master's graduate in Biology from the University of Gallifrey, coupled with my completion of the "Foundations of Teaching for Learning: Being a Teacher" and "Making sense of learning to teach" courses, I am keen to begin my teaching career at an institution known for its commitment to student development.

My education at the University of Gallifrey provided me with a profound understanding of biology, while my participation in courses on teaching equipped me with pedagogical skills and insights into effective teaching practices. My training emphasised student-centred learning and the development of engaging classroom environments. This combination of advanced subject knowledge and teaching methodology places me in a unique position to contribute to the science department at Springfield Academy.

Through these courses, I gained valuable insights into the intricacies of teaching, from lesson planning and reflective practice to the nuances of pedagogy and evaluation. These skills were further honed during my volunteer work at Springfield Community School, where I worked under Principal Charles Xavier, gaining practical experience in teaching and engaging students from diverse backgrounds.

I am particularly excited about the opportunity to bring my passion for biology to your students, fostering a love for the subject through innovative teaching strategies. My approach to teaching is rooted in creating an environment that is educational and supportive, enabling students to explore and appreciate the wonders of biology.

I am enthusiastic about the prospect of joining Springfield Academy and contributing to the legacy of fostering knowledgeable and socially responsible individuals. Enclosed is my resume, which provides further details about my academic and practical experiences. Thank you for considering my application.

Yours sincerely,

Brian Williams, MSc

SAMPLE CV

Brian Williams, MSc

Dedicated and passionate Biology Master's graduate from the University of Gallifrey, with comprehensive knowledge in biological sciences. I aim to leverage my academic background and teaching skills to inspire and educate students in a dynamic classroom environment.

EXPERIENCE

Teaching Assistant

University of Gallifrey, Department of Biology

September 2097 - May 2098

- Assisted in teaching biology to a class of 40 undergraduate students
- Developed and graded assignments and exams
- Monitored students during non-instructional time
- Maintained the established routine of the school

Volunteer Teacher

Springfield Community School, Springfield

January 2097 - June 2097

- Volunteered 10 hours per week
- Assisted with science classes for students aged 14-16
- Created and led 3 successful science workshops, enhancing students' practical skills and interest
- Maintained an orderly learning environment

EDUCATION

L. E. MAKAROFF

MSc in Biology

University of Gallifrey, Springfield

September 2096 - June 2018

- Graduated with Distinction (GPA: 3.9/4.0)
- Thesis: "The Impact of Environmental Changes on Marine Biodiversity"
- Received a research grant of £5,000

BSc in Biological Sciences

Springfield State University, Springfield

September 2092 - June 2096

- Graduated with First-Class Honours (GPA: 3.8/4.0)
- Final Year Project: "Genetic Analysis of Endangered Species" – Awarded Best Project in class

Certificates

- Foundations of Teaching for Learning: Being a Teacher
- Making sense of learning to teach

SKILLS

- Proficient in a range of laboratory techniques and tools
- Curriculum development and lesson planning
- Effective communicator, both verbally and in writing
- Skilled in student assessment and feedback
- Fluent in English and proficient in Spanish

~

SUMMARY

- A career in teaching is deeply fulfilling, offering the chance to shape future generations.
- It demands creativity, dedication, and a commitment to lifelong learning, both for oneself and for your students.
- The profession requires adaptability to meet the diverse needs of students and to navigate the ever-evolving landscape of education.

CHAPTER 16
TENURE-TRACK PROFESSOR

Knowledge will bring you the opportunity to make a difference.

CLAIRE FAGIN

So you've read through the other job descriptions, and you've decided that you want to stay in academia after all. Academics have the privilege to explore a wide range of topics, delving into subjects that fascinate them and contributing to the expansion of human knowledge. Professors often find themselves at the cutting edge of their respective fields, allowing them to make contributions and influence the direction of their discipline.

Academia offers numerous opportunities for global travel. Be it through attending international conferences or participating in research collaborations, academia frequently gives you the chance to pack your bags and embark on journeys that blend the essence of work with the spirit of wanderlust.

Another significant aspect of academia is the opportunity for mentorship and teaching. Professors play a crucial role in mentoring the next generation of students. You are a mentor shaping the minds and futures of the next generation of thinkers. Note that this might also involve students crying in your office about failed experiments and missed deadlines.

There can be collaboration in academia, offering ample opportunities to work with fellow researchers both domestically and internationally. When it works well, it can foster a sense of community and lead to groundbreaking collaborative research. The harmony of minds working in unison often results in discoveries that are greater than the sum of their parts. When it doesn't work so well, it can result in arguments about authorship ranking—a contentious issue that can fray the fabric of collaboration. Moreover, the intense pressures to secure funding and navigate the intricate politics of academic departments can sometimes lead to stressful situations that require careful navigation and emotional resilience.

Professors are under constant pressure to publish their research in prestigious journals and secure grants for their projects, a stressor commonly known as "publish or perish." The academic world is also highly competitive, with frequent rejections from journals, grant committees, and conference submissions. Such a competitive atmosphere can be demoralising and challenging to navigate.

Moreover, the workload for professors, especially those striving for tenure, can be overwhelming. Balancing research, teaching, administrative duties, and personal life can lead to stress and burnout. Academic positions, particularly tenured ones, are scarce and highly sought after, leading to a sense of uncertainty and instability for early-career academics. Until tenure is secured, professors often face job insecurity, with many capable academics not making the cut due to various institutional factors.

The responsibilities of a professor extend beyond teaching and research. They include developing and teaching courses, conducting original research, writing and publishing academic papers, guiding students, especially postgraduates, through their academic and professional journey, and participating in departmental and institutional governance. Additionally, professors engage with the wider community through public lectures, consultations, and collaborative projects.

The "two-body problem" in academia refers to the significant challenge faced by couples, particularly when both are seeking academic positions, such as professorships, in the same geographic area. Opportunities for tenure-track positions are limited and often scattered across different locations. For academic couples, this can mean having to choose between living apart for the sake of your individual careers or compromising on job opportunities to stay together. The "two-body problem" is exacerbated by the fact that academic institutions may not always have openings in both partners' fields at the same time, leading to complex negotiations and difficult decisions about prioritising career goals versus personal life. In my experience, navigating this challenge has involved a give-and-take approach, where my spouse and I have alternated in accepting the longer commute as a compromise with each step up the career ladder.

It's essential to highlight the importance of tailoring your CV and cover letter for academic positions. In academia, the degree of individual customisation required for these documents is significantly higher than in many other career pathways. When applying for academic positions, your CV and cover letter must reflect not only your qualifications and research accomplishments but also your fit within the specific department and institution. This involves a deep understanding of the department's research focus, teaching philosophy, and overall academic culture. Your application should articulate how your expertise, interests, and career trajectory align with the institution's goals and values.

Your cover should detail your research interests, teaching philosophy, and how these contribute to the department and the broader academic community. This level of personalisation ensures that you present yourself as not just a qualified candidate but as the right fit for the specific role and environment. Dedicate ample time and effort to customise your CV and cover letter for each academic application, as this can significantly impact your success in the competitive world of academia.

A career in academia is both challenging and rewarding. It requires a commitment to continuous learning, resilience in the face of rejection, and a passion for research and teaching. For those who value intellectual freedom and the pursuit of knowledge, academia can be a fulfilling career path.

ADVANTAGES TO BEING A PROFESSOR

- Freedom
- Being at the forefront of human knowledge
- Travel to conferences around the world
- Opportunity to mentor and teach
- Collaboration and networking

WHY LEAVE ACADEMIA?

- The pressure to publish and get grants
- Constant rejections can be tough
- It can be challenging to keep a healthy work/life balance
- Limited career opportunities
- Perception of limited job security at early stages

FILL THE GAPS IN YOUR CV ON YOUR OWN

- Develop skills like statistical analysis and programming
- Participate in professional societies
- Join the editorial team of a reputable journal

FILL THE GAPS IN YOUR CV IN YOUR WORK

- Publish substantive articles in peer-reviewed journals
- Write grant/fellowship applications
- Take on teaching assistant roles and guest lectures
- Contribute to departmental governance
- Mentor postgraduate students
- Supervise PhD or Masters candidates
- Engage in academic networking at conferences
- Collaborate on interdisciplinary research projects

GREEN FLAGS IN JOB DESCRIPTIONS

- Transparent and well-defined tenure track process
- Mentorship and support provided
- Faculty are encouraged to engage in sabbaticals
- "Collaboration, innovation, and dialogue are valued"

RED FLAGS IN JOB DESCRIPTIONS

- Tenure decisions are at the discretion of the department
- Heavy teaching load
- "Must consistently secure significant grant funding"
- "Individuals who thrive under pressure"

BEST COURSES TO TAKE

Teaching in university science laboratories

`coursera.org/learn/developing-university-lab-education`

This course was developed to improve the effectiveness of laboratory classes in higher education. It aims to support teachers in improving their teaching skills for active learning in university science laboratory courses. It will show you how laboratory sessions can differ concerning their aim and expected learning outcomes, how to engage students for learning, and how to cope with their different levels of pre-knowledge and experience and probe their understanding. Lastly, it will show how you could assess students in laboratory courses.

EMBO laboratory leadership

`lab-management.embo.org/`

This online course, spanning three days, focuses on enhancing leadership skills for researchers. Participants explore their leadership approaches, learn about team dynamics, and receive tools and techniques tailored to laboratory settings. The course, adapted for online delivery, features interactive modules with self-reflection and group activities facilitated by a diverse team of trainers experienced in life sciences leadership. To attend, participants need reliable internet, a webcam, a microphone, and the ability to use Zoom, with an emphasis on full participation in the interactive, seminar-style sessions.

~

SAMPLE COVER LETTER

Dear Dr Chin,

I am writing to apply for the Research Group Leader position in Synthetic Biology. Holding a PhD from the Mark Watney laboratory at California University, I am enthusiastic about the potential to lead innovative research within your institute.

During my postdoctoral fellowship at Unseen University, I focused on developing gene editing tools using CRISPR/Cas9 technology. This work has been instrumental in contributing to our understanding of gene regulation in disease. My approach has resulted in multiple publications, including a paper in *Nature Immunology*, where I was the first author.

I have secured over $500,000 in research funding from various sources. This, along with my experience in leading collaborative projects with multiple institutions, demonstrates my capability to advance research initiatives and manage a research group.

I am passionate about teaching the next generation of scientists. During my time at Unseen University, I taught undergraduate and graduate courses in biology and genetics. I actively participated in committee work, including contributions to curriculum development and research ethics.

My current research interests, which I wish to further at your institute, involve exploring the application of synthetic biology in developing targeted therapies for diseases. The synergy between my research aspirations and the programmes at your institute is strong. I am enclosing my CV, my current research interests, and a proposal for a future research programme.

Yours sincerely,

Sam Doe

~

SAMPLE CV

Sam Doe

A dedicated synthetic biology researcher with over 8 years of academic experience, holding a PhD in Biological Molecules from California University. Specialising in CRISPR/Cas9 technology, with a strong publication record including 12 peer-reviewed articles and demonstrated success in securing research funding exceeding $500,000. Sam's strong leadership skills are evident in team management, collaboration, and contributions to curriculum development and research ethics.

EXPERIENCE

Post-Doctoral Research Fellow

Unseen University

- Led a research team of 5 in developing novel CRISPR/Cas9 gene editing tools.
- Authored 6 papers published in high-impact journals, including Nature Immunology.
- Managed a research budget of over £300,000.
- Developed and taught undergraduate and graduate courses in synthetic biology and molecular genetics
- Supervised 3 PhD candidates, all successfully completing their dissertations
- Member of the Curriculum Development Committee, where I updated the synthetic biology curriculum and coordinated interdisciplinary course offerings with other departments

L. E. MAKAROFF

PhD Researcher

California University

- Conducted extensive research on synthetic biology, resulting in 6 publications.
- Collaborated with 4 international research teams.
- Presented findings at 15 international conferences
- Member of the Research Ethics Committee, where I reviewed and provided recommendations on research proposals

Education

- PhD in Biology, *Dupont University*
- Bachelor of Science in Biology, *Greendale Community College*

Key Publications

- "Innovations in CRISPR/Cas9 Gene Editing for Disease Treatment," Nature Immunology, Doe S., et al., 2098. (This publication received extensive coverage and analysis in a News and Views feature.)
- "A Novel Synthetic Biology Methodology in Immunology," Science Advances, Doe S. et al., 2099. (The data from this publication contributed to the filing of a patent)
- "CRISPR and the Future of Autoimmune Therapies," Journal of Molecular Biology, Doe S., et al., 2100. (This publication has been cited 181 times within a 12-month period)

Conference Presentations

- Keynote speaker at International Synthetic Biology Symposium, 2101
- Presenter at CRISPR Technology Conference, 2102
- Invited speaker at Gene Editing Innovations Conference, 2103
- Panelist at Molecular Biology Research Forum, 2104

Professional Affiliations

- Society for Synthetic Biology, Member since 2095
- International CRISPR Research Association, Member since 2098
- American Association for Gene Editing, Member since 2099
- European Society of Molecular Genetics, Member since 2100

Skills and Expertise

- Expert in synthetic biology with over 8 years of experience in CRISPR/Cas9 gene editing technology.
- Led research projects with teams of up to 10 members.
- Delivered 32 presentations at international conferences.

References

- Dr Jane Smith, Senior Researcher, Unseen University, jsmith@unseenuni.ac.uk
- Prof. John Black, Head of Biology Department, California University, jblack@cula.edu
- Dr. Emily White, Research Scientist, National Science Foundation, ewhite@nsf.gov

L. E. MAKAROFF

SUMMARY

- **Reasons to stay in academia include freedom, being at the forefront of knowledge, travel opportunities, mentorship and teaching, and collaboration and networking.**
- **Reasons to leave academia include pressure to publish and secure grants, constant rejections, work/life balance challenges, limited career opportunities, and job security concerns.**

160

PART THREE

STRATEGIES FOR CAREER TRANSITION

Career shifts often require us to step into unknown territories. This guide provides the tools and strategies to make your transition as smooth and successful as possible.

We'll start by breaking down the process into manageable tasks. You'll learn how to effectively plan and prepare for your career change, with practical examples, small steps, and real-life scenarios to help you actualise your career transition.

We'll explore the importance of gratitude and mindfulness in building resilience and maintaining a positive outlook, crucial for navigating the ups and downs of career change.

You'll find detailed guidance on preparing for interviews, including researching your interviewers, understanding the organisation's ethos, and practising your responses to common interview questions.

This part of the book addresses the challenges of relocating for a job and provides strategies for presenting yourself as the ideal candidate, regardless of your current location. It is your guide to transforming career transition challenges into opportunities for growth and development.

CHAPTER 17
HOW TO SHIFT CAREERS

The future belongs to those who believe in the beauty of their dreams.

ELEANOR ROOSEVELT

n this section, we explore a structured, step-by-step method to transition from one career or industry to another. Shifting careers might feel daunting, particularly when venturing into unfamiliar territory. It requires thoughtful planning and a strategic approach.

This guide breaks down the process into manageable tasks, providing practical examples of small, consistent steps to facilitate your career change. Discover how to effectively plan, prepare, and execute a successful transition, transforming challenges into opportunities for growth and development.

EVERY DAY

Thank you. Say "thank you" to a colleague who made your day brighter or easier. Not only will this lift your mood, but it will also be one step towards strengthening your professional network, which will be invaluable for gaining insights, advice, and potential job leads in your new career.

Be mindful of joy. Create a list of all the small experiences that make you feel a little bit happier. For example: "I like seeing the birds eating the bird seed I put out, I like laying on the grass, I like visiting the library, I like listening to podcasts in the morning, I like lavender scented shampoo..." Practising gratitude and focusing on joyful moments build emotional resilience, which is crucial for navigating the ups and downs of career change.

EVERY MONDAY

Fill a gap in your CV. Complete a module of an online class to help you gain expertise in the skills needed for your field. Look at the courses suggested in this book, as well as Coursera, EdX, Lynda, or your academic institution for access to courses. Simply by dedicating one hour every Monday, you can complete over 50 hours of online learning in over a year.

EVERY TUESDAY

Build relationships. Reach out to a professional in the field to learn more about the industry. Talk to people already working in the field. Ask an acquaintance if they would like to meet up for a coffee. Start making connections on LinkedIn and other social media networks for people in positions you would like to have.

EVERY WEDNESDAY

Intentional social media. By immersing yourself in a social media network filled with people from your chosen career, you will absorb the language, style, and hot topics being discussed, helping you appear well-informed during your application and interview. Follow people already working in your chosen career, and offer the occasional well-informed congratulations or comment on their posts.

EVERY THURSDAY

Celebrate success. Make a work-related social media post to thank someone who helped you, or to celebrate the success of someone you know. Tell a friend about a recent accomplishment and ask them what they are proud of achieving this week.

EVERY FRIDAY

Mark the end of the week. Note the end of the working week with a small ritual. Light a candle, buy a pastry, have a bath, watch an episode of your favourite TV show, sing your favourite song, inhale your favourite scent. Switching careers can bring uncertainties and anxieties; these rituals act as a counterbalance, offering moments of peace and enjoyment.

FIRST SATURDAY OF THE MONTH

Apply for a job. Read through the job advertisements and find a job that you want. Take the time to research the organisation, write a custom cover letter, and modify your CV specifically for that particular position, containing the buzzwords mentioned in the advertisement. Delete any courses or skills that are not relevant. Submit your application.

SECOND SATURDAY OF THE MONTH

Update your Master CV. Add any new publications, courses, events, seminars or awards to your Master CV and consider also adding them to your LinkedIn profile. Include the "who, what, when, where, why" of each course and a list of seminars you attend, with date, person, title and seminar program name. Even basic items like liquid nitrogen safety training or scientific writing courses can help boost your CV and demonstrate your passion for a particular career path. Ensure that your LinkedIn profile has a professional profile photo and banner that aligns with your career aspirations.

THIRD SATURDAY OF THE MONTH

Plan for the next month. Look through local university networking events, workshops, seminars, online activities, and career days. Look through Eventbrite and other online listings for professional events in your neighbourhood that spark your interests. Look outside your lab and immediate environment to encounter different career pathways. If you consistently find one set of events to be boring, that is helpful information; the career may be less appealing to you than you initially thought and can be replaced by other topics you may find surprisingly rewarding.

FOURTH SATURDAY OF THE MONTH

Schedule joy. Look at your list of daily pleasures, and use this to intentionally plan tiny activities or experiences throughout your next month to help you stay motivated as you switch careers. For example, find out the opening hours of your local library, look up free community events, schedule a walk with a friend, make time to write poetry, order your favourite snacks, note the date of the full moon, or download new podcasts.

~

SUMMARY

- Incorporate daily practices of gratitude and mindfulness to cultivate positivity and joy, which are essential for maintaining resilience and a positive outlook during the challenging process of switching careers
- Utilise specific days of the week for targeted actions, such as filling gaps in your CV, building professional relationships, leveraging social media for career immersion, and celebrating success.
- Allocate time each month to apply for jobs, update your CV, plan for future events and networking opportunities, and schedule enjoyable activities to enhance your personal life.

CHAPTER 18
THE JOB INTERVIEW

*The path from dreams to success does exist. May you
have the vision to find it, the courage to get on to it,
and the perseverance to follow it*

KALPANA CHAWLA

This chapter will guide you through the crucial steps of preparation, ensuring that when you sit across from your potential employers, you're not just another candidate, but <u>the</u> candidate. From mastering the art of first impressions to delving into the minds of your interviewers, I'll equip you with the tools to navigate this challenge.

BEFORE THE INTERVIEW

Ask for the names of everyone who will be in the room and look at photos of them. You want to use the people's names as often as possible. If you're unsure how to pronounce them, see if you can find a YouTube interview where their name is said. If it's a name from another country or culture and you can't find the pronunciation in a video, ask someone with the same background how to pronounce their first and last name.

Research these people in advance of the interview. What was their last tweet? What was the last item that they posted on LinkedIn? What was their previous publication? What does their biography say about them on the website? Use these questions to learn about issues that they care passionately about. If you also appear to know and care about the same topics, they're more likely to connect with you and hire you.

You should also read the organisation's annual report before stepping into that room and be prepared to speak about it during the interview. I remember hiring someone to produce the latest edition of a specific annual document. It was clearly stated in the job advertisement that this would be the person's primary role, but I found myself having to explain what this document was in the interview. This was not a good look for the person being interviewed. You need to learn the basics of the project that the interviewer is expecting you to step into before the interview takes place.

You need to practise for your interview in advance. You want to conduct at least three sets of two-hour interview practises — ideally twice that with multiple people. The simplest way to prepare is to ask a friend to read out the common interview questions found below, and then answer each question out loud. If you're feeling brave, you can also film your responses and then review them.

Interviewers typically start an interview by saying something like, "tell us about yourself". This is not something that you should wing. Memorise a 30-second elevator pitch about who you are, what you do, and why you are perfect for this job.

ON THE DAY

So it's your first interview. What do you wear? Whether it's a video conference or face-to-face interview, you should wear a suit: black and white, navy blue and white, or grey and white. Or a white shirt and blazer, or a nice dress or skirt and blouse. Please do not think you can attend an interview in jeans. This is not acceptable.

Turn up early and turn up with a copy of your CV, a notepad, and a pen. Have a copy of your CV for you and a copy for everyone in the room. If they are prepared, they might have copies printed out, but they will certainly not give you one. You want to make sure that you are looking at the same pieces of paper as they are. You want to take extensive notes during this interview because you will be referring to these notes when you write your reply email to them within 30 minutes of the interview ending. They are likely to make a decision that day about who to hire.

When you walk in the room, smile confidently and shake each hand (as long as it's pandemically appropriate) using the person's title and last name, such as Dr Smith and Professor Wang. If they invite you to use their first name instead of the title and last name, then feel free to do that.

If they offer you a glass of water or tea, you should accept it. In giving you a small courtesy, even if it's a glass of water, they are more likely to think of you in a positive light. You only need to take a sip, but it's a gesture of them giving you something that you then accept. This sets up a psychologically positive dynamic. If you start the interview with them offering you a gift and declining it, that interpersonal dynamic is slightly trickier.

If they start by asking a broad question like "tell us about yourself," you'll be prepared for your practice sessions. Show enthusiasm and be ready to discuss your professional and educational history. Your interviewer will ask about your previous roles, career goals, and skills. They'll also ask you to describe your leadership style. Be honest, but don't be negative. It's okay to discuss your weaknesses, but try to offset them with positive statements. For example, "I can be a bit of a perfectionist, but it means I always strive to do my best." Tell the interviewer about any relevant skills or experience and how these will help you excel in the role you're applying for.

Aim to answer all their questions in three short sentences. It's frustrating for the interviewers if you have one candidate who spends 10 minutes out of the 30-minute interview time telling them where they went to high school. With every sentence you speak, show that you will save these people time and improve their lives.

Look people in the eye and ensure equal eye contact with each person in the room, whether or not they are speaking. This can mean that people who are introverted or are not asking questions still feel included in the conversation.

Remember that every interaction is part of the interview. When dealing with the Personal Assistant who books your travel, remember that they have more time with the Director than you ever will. If you are invited to have lunch with potential colleagues, remember they report on you. Be pleasant to everyone.

While it may feel like they are interviewing you, you are also interviewing them. An interview is an excellent way to understand what it would be like to work at this organisation. It's tough to work in a toxic work environment; this is one way to determine whether it is a welcoming or hostile atmosphere. How do the people on your interview panel work together? Do they listen to each other? Do all of them have time to speak and ask questions? Or is there one person talking, and everyone else is listening? Do they listen to you? Do they give you time to answer your questions? Are they already pushing you to shorten your notice period or do an assignment for free before they hire you?

Listen to people and ask follow-up questions. Your colleagues want people who care about their work as well. It's easy to slip into prepared answers for questions that you think you've heard before. You must make sure that you listen to what they're asking.

Finally, remember to thank the interviewers for their time.

VIDEO CONFERENCE INTERVIEWS

If you are taking your interview via video conference, double-check that you've got the right time-zone and day.

The day before, make sure you have downloaded and updated the required video-conference software. Run through a test call with a friend using the same ideo-conference software.

Make sure that your background looks professional. If possible, don't use a virtual background, as this can make it look like you're trying to hide something. Keep your pets, children, and housemates out of the interview. Wear a blazer.

Ensure you have joined the virtual teleconference 10 minutes in advance to iron out any bugs like your computer suddenly needing an update.

Please don't take the interview on your phone. Using your computer looks more professional and shows that you take it seriously. If your computer is playing up and the phone is your only option because otherwise you'll be late, then use your phone.

You'll be tempted to look into the eyes of the people on your screen, but this is not what they see. They see someone who is not looking at them and who is looking a bit shifty, so make sure that you take some time whenever you're answering your questions to look directly into the camera, even if that puts you at a slight disadvantage because you cannot see the facial expressions of anyone on the video conference.

The organisation must trust that you can do professional video conferences for the hiring position. If you can conduct a professional discussion as part of your job interview, that puts you at a significant advantage.

IF THE JOB IS IN A NEW CITY

If you plan to move to the job location, clarify that in the interview. People are more likely to hire someone already in the area. They believe these people can hit the ground running and are more likely to accept the job. They think that these candidates are more likely to understand the culture and are more likely to be able to start earlier.

You are at a disadvantage if you apply from another city or country, so do your best to overcome this. In your cover letter, you should reassure them that you have everything taken care of for the move and that your move is imminent. You'll still be able to start as soon as anyone already living in the city.

You should also identify affiliations, networks, and collaborations with people or institutes in that new city. Completing an online course from a local university can demonstrate an affinity with the location. Mention these people and organisations to show that you already understand the area where the position is based.

～

EXAMPLE INTERVIEW QUESTIONS

Interview questions can seem deceptively simple or sound very broad, so it's important to be prepared with answers. Below, you'll find examples of what prospective employers want to hear when they're asking you some of the most common interview questions.

- Tell us about yourself
- Tell us what you know about our organisation
- Why do you want to work at our organisation?
- Why are you looking for a new job?
- What relevant experience do you have?
- What would your colleagues say about you?
- Where else have you applied?
- How are you when working under pressure?
- What motivates you to do a good job?
- What are your greatest strengths?
- What are your biggest weaknesses?
- Describe your biggest mistake and how you fixed it.
- Do you have any questions for us?

EXAMPLE INTERVIEW ANSWERS

Tell us about yourself

I completed a PhD in Gondor on white blood cells, supported by the Gondor Cancer Research Foundation. While I was living in Springfield, I attended a lecture by the Gates Foundation, which inspired me to move into Public Health. Currently I work at the International Health Coalition, where I manage a team and a budget responsible for report writing, advocacy, fundraising, and communications.

Why are you looking for a new job?

I am looking for a new challenge and would love the opportunity to do something extraordinary at an organisation like ABCD. When I saw this job posting, it sounded like an ideal match for my skill set and interests. Many of my family members have been affected by the disease.

Tell me what you know about ABCD

The ABCD is an umbrella organisation of 123 organisations in 46 countries. The ABCD has recently published a report on the 2124 Regulation. The ABCD is responsible for the annual ABCD Conference.

Why do you want to work at ABCD?

I have three main passions: Advocating for access to medicines, Empowering patients through social media, and Building stakeholder alliances. At ABCD, I would have the opportunity to utilise all three of these passions. I would also have the opportunity to work at an organisation committed to improving the lives of patients with disease.

What relevant experience do you have?

I have worked in public health and public affairs, with demonstrated success in changing policy and increasing disease awareness. I know how to bring together diverse organisations to reach a consensus. I know how to fundraise from the pharmaceutical industry, medical devices, and high net worth individuals.

What would your colleagues say about you?

I am hard-working - I won't give up until the report is final. I am detail-oriented - I enjoy pouring over budgets and action plans to ensure that everything looks just right. I am a good manager - I work well with others, ensuring that my team feels supported and empowered.

Where else have you applied?

I have also applied for a job at the WHO.

or

I've only applied for this position so far since it was such a perfect fit for me.

How are you when you're working under pressure?

At my current job, our deadlines are set in stone and I excel at meeting them. I work with my team to ensure that everyone knows all the tasks that must be completed on time, and I meet with them regularly to identify any roadblocks. I enjoy working under pressure to deliver results.

What motivates you to do a good job?

Bringing people together to advocate for access to medicines. Being able to celebrate successes with my team.

What are your greatest strengths?

Motivating a team, fundraising, and community building.

What are your biggest weaknesses?

When I began managing a team, I got very attached to all the projects. As I have been given more responsibilities, I have learned to delegate, trust, and assess. When I was new in the organisation, I received criticisms from members of the International Health Coalition. I now know not to take these criticisms personally and to be thankful we have such passionate members. It can be challenging to receive constructive criticism. However, I now understand that feedback is a gift.

Describe your biggest mistake

Early on, as a manager, I was working with a new team member on writing a report. We met once a week, and each time I would ask him if he was having any problems, he would say "no". When I finally saw some of his results, they were blatantly wrong. My mistake was that I had failed to know that he was struggling and he had difficulties communicating this. Now I ensure that the first job of all new team members is to have their work checked by an experienced team member.

Do you have any questions?

- What is the biggest challenge facing ABCD now?
- Please tell me more about the culture and workplace of ABCD.
- What qualities does ABCD look for in a new hire?
- What would you be looking for the new hire to achieve in the first six months?

SAMPLE FOLLOW-UP EMAIL

Dear Prof L, Ms A, and Ms M,

I had a wonderful afternoon yesterday talking with you all about the European Federation. I am very impressed by this energetic organisation's achievements and your passion and knowledge of the challenges ahead. Your action plan for the year is a well-organised balance of advocacy, research, and capacity building.

I am an excellent match for the role. I have experience in team management, project management and research. I am skilled at low-cost website development and developing online platforms. I am willing to work to ensure that ABCD continues to grow and delivers meaningful improvements in the lives of patients.

If you have further questions, please don't hesitate to contact me.

Best regards,

Sam.

~

SUMMARY

- **Prepare for the interview by researching the interviewers and the organisation, including their recent social media activity and publications.**
- **Practise your interview responses in advance, aiming for concise and impactful answers that showcase your relevant skills and experiences.**
- **If you're considering relocating for the job, communicate your readiness and highlight any connections or affiliations you have in the new city to demonstrate your familiarity and commitment to the location.**

CHAPTER 19
SURVIVING YOUR CURRENT ROLE

*You may encounter many defeats, but you must not be
defeated. In fact, it may be necessary to encounter
the defeats so you can know who you are, what you
can rise from, and how you can still come out of it.*

MAYA ANGELOU

Graduate work is an immensely challenging journey. It demands your mental acuity and a significant investment of your emotional and physical well-being. Recognising this, it's crucial to approach this period with a holistic self-care strategy. This means paying attention to your physical and mental health, cultivating relationships, and creating a nurturing work environment.

Regular exercise, maintaining a healthy diet, ensuring adequate sleep, and integrating self-care activities such as meditation or yoga are foundational to physical well-being. These practices keep you physically fit and bolster your mental resilience.

Mental health is equally important. It's essential to take regular breaks, engage in social activities, and, if necessary, seek professional support. Learning to appreciate your unique strengths, avoiding unhelpful comparisons, and maintaining a positive outlook are key elements in safeguarding your mental well-being.

In the following sections, you will find specific strategies and tips to help you survive your graduate work.

TAKE CARE OF YOUR PHYSICAL HEALTH

To survive your current PhD or post-doc, taking care of your physical health is essential. Create a routine that allows you to have time to exercise, eat healthily, and get at least 7 hours of sleep every night.

Additionally, it is important to prioritise self-care activities such as baths, gardening, meditation, or yoga, which can help reduce stress and increase focus. Limit caffeine intake, stand up once an hour, drink water, and take vitamin D if you live in a place with limited sunlight.

TAKE CARE OF YOUR MENTAL HEALTH

Taking regular breaks throughout the day and engaging in social activities help boost morale and provide a sense of connection.

If times get tough and aren't improving, consider speaking to a doctor about whether or not a prescription for an antidepressant or anti-anxiety medication would work for you. Take advantage of free counselling if your institute offers it. It's always good to learn coping techniques and understand your mind better.

Don't compare yourself to others. Everyone is unique and has their own fantastic set of skills and talents. Trying to measure yourself up to others will only lead to feelings of inadequacy and jealousy. Switch off your social media. Instead, focus on your strengths and abilities and work on developing them.

Write down one thing that you are grateful for every day. For example, I appreciate my cat, heating, trains, baths, and family.

CULTIVATE YOUR RELATIONSHIPS

Finding ways to stay connected with friends and family and maintain an active social life is essential. Who are the people you will mention in the acknowledgements section of your thesis?

Cultivate the relationships with your loved ones by creating simple yet regular rituals. Aim to have a meal or coffee together daily or weekly with someone who makes you smile. Plan a fun activity with family or friends every week; it could be cooking a meal together, watching a TV show, playing a board game, or going for a walk. Once a year, go on a rejuvenating holiday with your favourite people.

TAKE IT HOUR BY HOUR

I know that some days can be extremely tough. There have been some work days when it felt like an accomplishment to get through each hour. On those days, I drew four lines on paper to create nine boxes and celebrated surviving each hour with a tick. It was a small reminder that I was doing my best and making progress.

Sometimes it takes strength to mentally pull yourself out of a dark place. Our brains can tell us that our lives are going to be like this forever, and nothing will ever change. We all have tough days. Still, finding ways to celebrate our accomplishments is crucial, no matter how small.

YOUR WORKING SPACE

Whether you work from home or an office, make yourself a pleasant working space. Choose a small potted succulent and an ornament to brighten up your room. A photo of something that makes you smile, like your pet or your favourite holiday spot, can also help inspire positive feelings while working.

Position your chair to look out a window. Set up your work area with comfortable seating and good lighting, and make sure you have all the necessary supplies and equipment close at hand.

Ask your institution to purchase a secondary monitor and learn how to set it up as an extended rather than a cloned one. Choose a good mug for tea and a pretty glass for water, and have a blanket nearby for those chilly days. Keep your phone and other devices out of reach so you don't get distracted by them.

~

SUMMARY

- Prioritise your physical health by establishing a routine that includes exercise, healthy eating, and sufficient sleep.
- Avoid comparing yourself to others and focus on your own strengths and abilities.
- Remember that tough days are normal, and finding ways to celebrate your achievements, no matter how small, is important. And asking for help is not weak.

~

CHAPTER 20
BUILDING YOUR CAREER TOOLKIT

You are going to have an extraordinary life. You are good enough. Be brave.

LE MAKAROFF

TIPS FOR COVER LETTERS, CVS AND REFERENCES

There is no quick way to write a cover letter and CV. Unfortunately, every time you make a new job application, you will have to spend many hours researching the job, the person you're writing to and the institution you want to work at. Find out about the organisation, their values, and what they stand for.

You might think it's more efficient to spam the same generic CV and cover letter to 100 people, but trust me, we know when you are doing this, and the application goes straight into the trash. We want to see that you are interested in our organisation and have put some effort into applying to us, so tailor your CV and cover letter for that specific role.

Always ask someone else to proofread your cover letter and CV before you send them off. This will ensure that there are no mistakes that could cost you the job.

How do employers read a CV and cover letter?

As someone who has hired many employees, I personally first look at the CV and the job title that the person currently holds. Does it match the job title of the position I am hiring for, or does it seem like a reasonable next step in their career? Often, people apply for roles to advance when they've reached the growth limits in their current roles. If the job title and a list of responsibilities are similar to the job advertisement I am considering, I will review it in more detail. If there's no direct match, I'll briefly glance at previous jobs and the cover letter. However, I allocate time for the cover letter, particularly if the applicant wants to explain why they are not such a mismatch. A well-crafted cover letter allows candidates to articulate why they are switching jobs. They might acknowledge that their current job doesn't directly align with what I'm looking for, but they can demonstrate in their letter that they possess the attitude and perseverance necessary to learn and grow in the new role. This approach shows a thoughtful understanding of their career trajectory and the potential they bring to the new position.

Remember that the person reading the CV and cover letter often has at least 100 of these to get through and shortlist. Often the recruiter has an idea of this mythical perfect match. Your job is to convince them that you are as close as possible to that mythical person. Use the language that was in the job description. Don't worry that it sounds too similar to the job advertisement. Don't try to be too innovative. Simply explain that you have the skills that they are looking for. List out the skills they have said are essential and then give evidence for how you match them.

Sample references

Is your potential employer asking you for a written reference from one of your referees? There's no harm in drafting the first version yourself and sending it to your referee. After all, you might remember more than them about how you helped them.

~

I worked with <<Dr your name>> for <<number>> years in my role as <<referees' role>> at the <<name of organisation>>. <<She/he/they>> <was/were>> responsible for projects including the <<list projects>>.

~

<<Dr your name>> is skilled in business development, working very well with new and existing clients, determining their needs, and developing attractive contracts for both parties. <<She/he/they>> <<has/have>> a strong understanding of the research capabilities of the organisation and can utilise that knowledge to deliver high-quality projects with realistic scope, budget, and timelines.

~

<<Dr your name>> has demonstrated excellent leadership and managerial skills. <<She/he/they>> manage/s a team that includes interns and placement students. I have witnessed <<his/her/their>> excellent mentoring abilities and a focus on ensuring the well-being of the team.

~

<<She/he/they>> <<has/have>> also has experience working with vendors, consultants, and partners. <<She/he/they>> have an extensive personal network in private industry, academia, and non-government institutions. She has also increased the visibility of the <<organisation>> through <<his/her/their>> participation in congresses, expert meetings, and webinars. <<Dr your name>> has excellent written and verbal communication skills and is very enthusiastic about research and public policy.

~

I have no hesitation in recommending <<him/her/them>> to you and am sure <<she/he/they>> would be an asset to any organisation.

SOCIAL MEDIA BEST PRACTICES

Facebook, TikTok, and Pinterest

Your employer will research you. Check that your social media accounts like Facebook, Snapchat, Instagram, TikTok, and Pinterest are either set to private or only show you in a light that you would be proud for the CEO of your new organisation to show at their annual meeting.

LinkedIn

A LinkedIn account is a necessity. It's a great way to connect with potential employers and fellow professionals. Check that your LinkedIn account is public and shows you in the best possible light. Delete any controversial posts.

Ask some co-workers to write you a LinkedIn reference, which can act as a living source of referrals from people who have worked with you and can attest to your skills and strengths. You can offer to give them ideas for things you'd like them to highlight, both to make it easier on your reference and to ensure they mention what's important to you.

Example LinkedIn references:

66 Mohammed is enthusiastic, creative, and has extensive knowledge of the European health network. During his time at the European Federation, I promoted him from Public Affairs Coordinator to Head of United Nations Affairs. I was particularly impressed by Mohammed's extensive understanding of the health policy landscape and his ability to build relationships and generate new ideas for collaborations. Mohammed would be a true asset for any positions requiring engagement and advocacy.

66 Annetta is very talented in translating global evidence into concrete policy recommendations and using this to successfully advocate for change among stakeholders such as the United Nations and the World Health Organization. Her detailed knowledge of the advocacy process, strong strategic thinking skills, and detailed analysis of relevant issues make her an asset to the International Health Foundation.

66 I've worked with Maria at the International Health Coalition for over two years. During this time, she has coordinated the Global Health Scorecard, prepared social media advocacy campaigns, and called for action on health during the G7, G20, and SDG meetings. I have seen her develop her global

advocacy, public speaking, and English communication skills. She is dedicated, enthusiastic, and flexible.

I worked with Sam for over a year at ABC. She worked tirelessly to drive the charity's policy, awareness, and patient support work forward. She made a considerable effort to make me feel part of the team as a part-time freelancer. Her knowledge of health and experience with public policy on international stages gave the charity real credibility and a voice to make change.

I worked very closely with Joan for almost two years. While she was dealing with many projects simultaneously, all projects were delivered on time and were very successful due to her management, professional and personal skills, and excellent discipline in meeting deadlines. Moreover, she set and shared high-quality scientific standards. She was irreplaceable in our team, with her comprehensive summarising, writing and editing abilities, and analytical perspective. She personally and professionally contributed a lot to each team member's performance.

Even in the busiest periods, she consistently made herself available to clearly explain any task. Beyond this, Sam has solid project management skills. She can focus on the intricate details of a project at any given moment while never losing sight of the whole and longer-term picture. In addition to her apparent technical expertise, Sam is gifted in building professional and productive relationships with partners and collaborators. This combination of qualities is invaluable in any team and organisation, and the extent to which Nneka demonstrates them is unique.

" Few people have the chance to report to a great mentor, but I did when working in Sam's team at the International Health Coalition. Sam's ability to motivate a team to care about its projects and be invested in their success is impressive. She always has a creative, positive outlook and is good at organising and bringing people together. Sam has the quality of being a friendly and patient person who works well and professionally with those around her. Sam is a person you go to for conflict resolution and problem-solving; she was always able to find compromises that satisfied all involved parties, either external or internal. Any employee would be lucky to have Sam as a manager.

LinkedIn summary

A well-crafted LinkedIn summary is crucial.

It's your opportunity to make a solid first impression to recruiters, colleagues, and industry connections. A good summary succinctly showcases your background, experience, and professional achievements. It should highlight your unique skills and accomplishments, give us a glimpse of your career aspirations, and demonstrate your enthusiasm for change.

Your LinkedIn summary should start with a strong opening that captures attention. It might include a powerful statement or a concise overview of your key strengths or professional ethos. From there, it should succinctly showcase your background and experience, highlighting your major professional achievements and skills. Instead of simply enumerating your job titles, focus on what you've accomplished in those roles and how they define your professional identity.

Incorporate your unique skills and accomplishments, ensuring they align with the career path you are pursuing. Whether it's leading successful projects, driving innovation, or fostering collaborative environments, your summary should illustrate these abilities with concrete examples.

Your career aspirations are equally important. This is your chance to share where you hope your career will take you, aligning your past experiences with future goals. This section of the summary should exude enthusiasm and readiness for new challenges, showing your commitment to growth and learning.

A personal touch can make your summary more relatable. Consider including a brief mention of what motivates or inspires you professionally. This could be a commitment to continuous learning, a passion for a particular aspect of your industry, or a drive to contribute to meaningful change.

Your LinkedIn summary is an opportunity to demonstrate your communication skills. It should be well-written, clear, and engaging.

Here are some examples:

~

I am a native English-speaking manager with ten years of international experience working in Gondor, Eddis, and Europe. I supervise a diverse global team of researchers, analysts, statisticians, and health economists.

~

I have a broad knowledge of economics, epidemiology, and public health impact of non-communicable diseases. In the past year, I have supervised the production of the International Health Coalition Health Atlas and the Health Guide, launched a major website, produced high-impact scientific papers, presented symposia at international congresses, and written for Health Voice.

∽

Based in Pyrus, I am skilled in business development and have worked with partners in industry, academia, United Nations institutions, and non-government organisations.

∽

I am passionate about health and development. Neglected tropical disease interventions are cost-effective ways of saving lives. Cancer has an immense impact on health, society, and our economy. Timely diagnosis and treatment of bladder disease allow many people to live longer and better.

∽

I am responsible for managing diverse teams, producing policy papers, designing multi-year strategies, fundraising, launching significant projects, presenting at international congresses, and writing articles for journals and magazines. I am skilled in business development and have worked with partners in industry, academia, international institutions, and non-government organisations. I have a broad knowledge of economics, epidemiology, and the public health impact of non-communicable diseases.

∽

I'm fascinated by relationship management and enjoy meeting like-minded people. Feel free to contact me directly if you share my interests.

~

X (Twitter) / Mastodon

A professional X (Twitter) / Mastodon account can be beneficial for your career. This X (Twitter) / Mastodon account should include posts about your research and other interesting papers and presentations. Learning to summarise a paper in 280 characters is a helpful skill. It is also an excellent way to network with other researchers and academics in your field. X (Twitter) / Mastodon can be used to connect with potential collaborators, discuss research ideas, and stay up to date with the latest developments in your area.

Many employers use X (Twitter) to advertise job openings and to promote their company or organisation. By following potential employers on X (Twitter), you can stay up to date with their latest news and developments, and you may even be able to find out about job openings before they are advertised elsewhere.

Sample X (Twitter) or Mastodon biographies

they/them. President @ABCDgroup. Patient advocate ▬ & ◕ citizen. Views are mine.

she/her. @ELSDKJF board. @SDKFJ member. Spearheading the #hashtag campaign.

he/him. Health, sustainability, circular economy, climate, digital advocacy, media ⁺ ◦ Opinions mine.

Medical researcher. ⚗ ✒ at @SLDKFJ. @KDF volunteer.

Co-founder of @LSKDFJ /PhD candidate / 🏉 lover/ co-host of @ABCD_Podcast

Eager to empower citizens; PhD Candidate at @Unseen_University

PhD candidate | Public Health & Biostats at @Unseen_University. @YoungActivities alumnus

Academic. Researcher in the Field of Industrial Relations. PhD Candidate 🎓 | | Sunnydale College Alumnus 🏆 | Genovia Raised 🏙

≈

SUMMARY

- Dedicate time to customising your CV and cover letter for each application, ensuring they reflect your knowledge of the organisation and role, and having your references and documents proofread by others.
- Employers look for relevant job titles and responsibilities in CVs and appreciate well-crafted cover letters that clearly explain career transitions or mismatches, usually spending limited time on each application.
- Maintain a professional presence on platforms like Facebook, Instagram, TikTok, Pinterest, and LinkedIn, and use Twitter or Mastodon for networking and showcasing your expertise in your field.

AFTERWORD

Leave a mark in this world. Have a meaningful life, whatever definition it has for you. Go towards it. The place we are leaving is a beautiful playground where everything is possible. Yet, we are not here forever. Our life is a short spark in this lovely little planet that flies incredibly fast to the endless darkness of the unknown universe. So, enjoy your time here with passion. Make it interesting. Make it count!

- U/MYLASTTIE ON REDDIT

The journey from an academic setting to a new career path is exciting and daunting. This guide has journeyed through various career possibilities, encompassing roles in both small and large charities, museums, civil service, and the sphere of health and safety. It has also shed light on more specialised positions, such as technicians, medical writers, journalists, and opportunities in the dynamic realms of the pharmaceutical industry and biotech start-ups.

Additionally, I've provided insights on how to shift careers effectively, ace job interviews, and navigate the complexities of your current role.

Remember, it's perfectly fine if, after assessing your needs and aspirations, you decide to continue your journey within academia, just as valid as exploring new professional horizons.

This guide is meant to illuminate paths and empower you with knowledge and tools, whether you choose to branch out into a new field or deepen your roots in the academic world.

Your career journey is as unique as you are. Every step and every decision is part of a larger journey towards fulfilling your professional potential. Ultimately, it is about discovering and achieving your version of happiness.

Life's too short to spend our time on work that doesn't spark joy in our hearts. It's tough to find success in something that fails to light up our eyes. After all, patience, passion, and dedication flow naturally when we're doing something we truly love.

So here's my call to action for us: Let's grab the reins of our lives. Let's own our choices and their outcomes. Our lives should be sculpted by the choices we bravely make, not by the opportunities we let slip by. Let's live by design, not by default.

〜

As a first-time self-published author, I'm truly grateful for your time spent with my book. If my work has resonated with you, kindly consider leaving a review. Your feedback not only supports my writing but also helps others discover this book.

Thank you immensely.

〜

ACKNOWLEDGMENTS

Freya Marske once noted that one can tell a debut book from the length of the acknowledgements. I wholeheartedly endorse this tradition.

This book owes its existence to a constellation of remarkable individuals and elements.

To the plants that sat beside me as I type this, for your oxygen and your reminder that progress often happens unobserved.

To the bees, for their bumbling.

To the jackdaws, the wood pigeons, and the magpie gang for bringing playfulness, cunning, and wildness to my everyday life.

To the botanic gardens for your acacias, banskias, and wollemis and for being a serene haven.

To the technology that has enabled me to reach further, learn faster, connect deeper, and hear more clearly.

To the sea, for its vast mystery and reminding me of opportunities beyond the horizon.

To the Winter solstice, the Spring equinox, the Summer solstice, and the Autumn solstice, for your enduring rhythms.

To the night sky, for showing me that my existence is insignificant in the best way.

To the intrepid explorers Sojourner, Spirit, Opportunity, Curiosity, Perseverance, and Zhùróng, for sharing your adventures in a realm far from home.

To the authors Naomi Novik and Martha Wells, for bringing El and Murderbot into my life, and for making them feel like friends.

To the artists Charley, Beth McCarthy, Maisie Peters, Kitty Perrin, Peach PRC, and Taylor Swift, whose music is a wellspring of energy, even though copyright legalities necessitated the removal of your lyrics from the final draft of this manuscript.

To Emily Starbuck Gerson, for not only your wonderful friendship and virtual Beat Saber battles but also your astute eye and incredible 605 suggestions that sculpted this manuscript into its current form. Any remaining errors are solely mine.

To Michaël and Cedric Van Dorpe, for keeping me abreast of all trends in technology and inspiring a life of optimisation. Alex Filicevas, for your comprehensive feedback. The three of you led me to realise that this manuscript was more than one book.

To Siobhan Hall, for your sharp analysis of the journalism chapter. To Anna Laštůvková, for your meticulous review. To Jess Beagley, for your unique perspective on charity work. To Michelle Marchione, your presence in the pharmaceutical world is inspiring.

To Amy Nicholls and Ewan Lawrie - even though we only share 25% of our DNA, I am 100% proud to be your sibling.

To Cathy Lippmeier, for graciously sharing her birthdate with me, always knowing the right questions to ask, and for her profound interest in the answers.

To Adam Bales, for soothing pasta, conversations that drive my aspirations, and understanding of my threat system.

To my PhD peers - Amy Jennison, Katharine Gosling, Fleur Roberts, Rob Stagg, Angelo Theodoratos, and Anton Wasson - your camaraderie turned a daunting experience into something that I almost remember with fondness.

To Karen Coulter, Bess Fitzgibbon, Sayuri Kusumoto, Joyce Man, Yuan Pan, Siew-Yit Samshi, Nneka "Peach" Smith, Charlie Stokes, and Liz Walsh, for being rainbows in my life, casting a kaleidoscope of light.

To James Dooley for summer BBQs and winter laughter.

To Emma Beddington, Susan Beale, and Henry Scowcroft, your journeys from concept to bookshelf have been guiding stars.

To Elisheva Fox, I am so proud to watch you flourish as a poet and a person, even from beyond the ocean.

To the NaNoWriMo and self-publishing community, your collective spirit and advice have been a tremendous resource.

To my mother, Sonya, thank you for your belief in me. To my late Nana, Gwen, I hope that your literary legacy lives on here.

To Adrian and Hayden Liston, your support has been the heartbeat of this journey. To Mint, our adorable cat, your gentle snores have been a delight during the long hours of writing. To Pepper, I miss you - your pawprints have made a permanent mark on my heart.

This book is a tribute to all who, in various ways, have contributed to its creation.

∽

ABOUT THE AUTHOR

Dr Makaroff's journey in the life sciences led to a PhD in Medical Research and a Master of Public Health. Working across various roles, from a Chief Executive at a national charity to a Health Outcomes Manager at a global pharmaceutical company, Dr Makaroff has delved deep into both the academic and practical sides of science.

Dr Makaroff has been honoured with numerous awards, including the European Health Award and the Communique Award for Excellence in Communication.

Beyond professional pursuits, Dr Makaroff finds joy in watching rocket launches and Mars rovers, delving into the worlds of local libraries and virtual reality, all while sharing a home with a supportive spouse, a beloved child, and a cherished cat.

For more books and updates:
`www.makaroff.ink`

ABOUT THE FONT

The Palatino font, designed by Hermann Zapf in 1948, is a serif typeface inspired by the calligraphy and type designs of the Italian Renaissance. It was named after the 16th-century Italian master of calligraphy, Giambattista Palatino, and is a font that offers both a beautiful form and practical functionality.

The typeface strikes a balance between calligraphic flair and understated formality. Its strokes emulate the movements of a pen on paper, adding a humanist touch to its appearance.

It has large open counters (open spaces within letters), contributing to its readability. The characters have a generous x-height (height of the lowercase letters), enhancing legibility. The font has strong, legible serifs, which lead the eye along the line of type, making reading smoother.

In 1984, Palatino was among the first typefaces to undergo digitisation, making it accessible on early personal computers. This transition was due to the pioneering work of typographer Gudrun Zapf von Hesse, who demonstrated great expertise in crafting bitmaps for various alphabet designs.

www.ingramcontent.com/pod-product-compliance
Lightning Source LLC
Chambersburg PA
CBHW011845200326
41597CB00028B/4711